Multiple Choice Qu

Multiple Choice Questions in Physiology

Sheila Jennett MD, PhD, MRCP (GLASG)
Reader in Physiology, University of Glasgow

Oliver Holmes MB, BS (LOND), MRCS (ENG), LRCP (LOND)
Senior Lecturer in Physiology, University of Glasgow

Churchill Livingstone

EDINBURGH LONDON MELBOURNE AND NEW YORK 1986

CHURCHILL LIVINGSTONE
Medical Division of Longman Group UK Limited

Distributed in the United States of America by
Churchill Livingstone Inc., 650 Avenue of the Americas,
New York, N.Y. 10011, and by associated companies,
branches and representatives throughout the world.

First published 1983 (Pitman Publishing Ltd)
Reprinted 1984
Reprinted 1986 (Churchill Livingstone)
Reprinted 1989
Reprinted 1990
Reprinted 1991
Reprinted 1993

ISBN 0 443 03620 9

British Library Cataloguing in Publication Data
A catalogue record for this book is available from the
British Library

Library of Congress Cataloguing in Publication Data
Jennett, Sheila.
Multiple choice questions in physiology.

Includes index.
1. Human physiology—Examinations, questions, etc.
I. Holmes, Oliver. II. Title. [DNLM: 1. Physiology—
Examination questions. QT 18 J54m]
QP40.J47 1983 612′.0076 83–11451

Produced by Longman Singapore Publishers Pte Ltd
Printed in Singapore

Contents

Introduction

These yes/no questions are designed for preclinical medical students and postgraduates preparing for the Primary Fellowship examinations. The questions have been extensively tried out and modified in the process of their use in examinations and self-assessment papers for medical students. We are indebted to our students and our colleagues in the Department of Physiology for pointing out ambiguities and suggesting improvements. Nevertheless, comments on remaining imperfections will be welcomed.

The questions in this book are arranged under topics or systems; there is necessarily some overlap and particular items are indexed. The questions assess aspects of a candidate's grasp of physiology by testing his ability to solve numerical problems (e.g. Q 6–18), make valid inferences from data (e.g. Q 2–26) and recall factual information (e.g. Q 1–05). Questions of this nature cannot test the candidate's ability to express himself linguistically or to construct logical arguments. Nor can they test originality or creativity.

The five items in each question are concerned with a topic or concept indicated in the initial phrase or 'stem'. Each of the five items reads on from the stem to make a complete statement which is either correct or incorrect. We have tried to set questions such that the decision for each statement is entirely separate from the decision about any other statement.

Seldom can ideal wording be constructed; we often have to compromise between the acceptability of common usage with clarity on the one hand and pedantic precision with absolute unambiguity on the other. With these limitations, each statement is designed to seem plausible to those with inadequate knowledge, yet at the same time to be unequivocably right or wrong.

Advice to candidates

In any one question, read the stem and item (a) and decide 'Yes' or 'No'. Then read the stem and item (b) and decide 'Yes' or 'No' and so on. The items are designed to be separate, so consider them independently.

A candidate who guesses randomly, without any basis of knowledge, should in all fairness be left with a mark of zero. The commonest marking system deals with this by giving +1 for a correct answer, −1 for an incorrect answer and no mark for a blank or 'don't know' answer. In medicine, intelligent guessing without secure information may sometimes be appropriate and useful. A wild guess, however, is worse than an acknowledgement of ignorance. Likewise in attempting to answer this type of question, answer 'Yes' or 'No' if you are sure, or if your choice is reasonably securely based but leave your answer blank if it would be pure guesswork.

Construction of questions

In the construction of these questions there are some particularly common pitfalls which we have tried to avoid:

Statements must not be mutually exclusive. If for example the answer to item (a) is 'Yes', this must not mean that the answer to any other would necessarily be 'No'.

Each part of a question should test only one bit of knowledge or deduction. The following example illustrates how questions should not be constructed.

Example:

A rise in arterial blood pressure

(a) leads to a decrease in heart rate because of increased baroreceptor activity.

(b) causes cerebral vasoconstriction which reduces cerebral blood flow.

Although both things in (a) are right, it is not appropriate to test both the fact ('decrease in heart rate') and the mechanism ('increased baroreceptor activity'). (b) has the same fault but is also treacherous because the first part of the statement is correct and the second is not.

The items tested in the example should have been separated into four questions.

Questions

Section 1 Digestion, Absorption and Metabolism

Q 1-01 Carbohydrates:

(a) start to be digested in the stomach.
(b) are ultimately broken down at the brush border of intestinal mucosal cells.
(c) pass into the blood as monosaccharides.
(d) after digestion, can be absorbed only down a diffusion gradient (i.e. only when the intestinal luminal concentration is higher than plasma concentration).
(e) have about the same calorific value weight for weight as fat.

Q 1-02 With reference to gastric function:

(a) the average meal has moved on from the stomach after half an hour.
(b) contractility is augmented by sympathetic stimulation.
(c) when the quantity of ingested material increases, the intragastric pressure increases.
(d) the longitudinal muscle coat has a basic electrical rhythm.
(e) histamine inhibits the production of gastric acid.

Q 1-03 On the left is a list of gastric secretions. Are the items on the right appropriately associated with each secretion?

(a) Hydrochloric acid	Increase in bicarbonate in gastric venous blood
(b) Hydrochloric acid	Necessary to life
(c) Pepsinogen	Remains inactive above pH 6
(d) Intrinsic factor	Another name for vitamin B_{12}
(e) Gastrin	Secretion stimulated by vagal activity

Q 1-04 Concerning gastric secretion in a normal adult:

(a) the lowest pH which can be attained in the stomach is about 4.5.
(b) the histamine receptors in the stomach can be pharmacologically stimulated without significant stimulation of histamine receptors in the lungs.
(c) pentagastrin injection can cause a maximal secretion of acid by the stomach.
(d) excessive acid secretion is prevented by an effect originating in antral receptors.
(e) gastrin comes mainly from cells in the fundus of the stomach.

Q 1-05 Concerning gastro-intestinal secretions:

(a) gastro-intestinal hormones are steroids.
(b) secretin is secreted from the pancreas.
(c) secretin stimulates flow of pancreatic secretion with a high concentration of bicarbonate ions.
(d) the maximal flow of pancreatic juice in response to cholecystokinin and secretin given together is greater than the maximum produced by either hormone alone.
(e) the pancreas secretes more than one litre of juice per 24 hours.

Q 1-06 In the normal human:

(a) cholecystokinin is the most important hormone concerned in the neutralization in the small bowel of acid from the stomach.
(b) in the gall bladder, chloride ions are secreted into the bile.
(c) the emulsification of dietary lipid by bile salts assists intestinal absorption of lipid.
(d) at least 95% of bile pigments secreted by the liver are reabsorbed in the gut.
(e) most of the water absorbed in the intestinal tract is directly derived from dietary intake.

Q 1-07 Concerning pancreatic secretion:

(a) the pancreatic juice secreted in response to vagal stimulation is rich in enzymes.
(b) secretin can produce a greater flow of pancreatic juice than the maximal flow in response to vagal stimulation.
(c) atropine blocks the secretogogue effects of vagal stimulation.
(d) atropine blocks the secretogogue effects of secretin.

(e) secretin causes the production of a pancreatic secretion which is more alkaline than that secreted in response to CCK (cholecystokinin).

Q 1-08 Concerning fat digestion and absorption:
(a) in a healthy human, 5% is the upper limit of ingested fat which appears in the faeces.

Malabsorption of lipids is likely to occur as a result of:
(b) hepatocellular damage.
(c) obstructive jaundice.
(d) failure of pancreatic secretion.
(e) removal of the gastric antrum in a partial gastrectomy.

Q 1-09 Bilirubin released into the circulation from the reticulo-endothelial system:
(a) is bound by plasma proteins.
(b) is conjugated with glucuronic acid in plasma.
(c) contains ferric ions.
(d) is readily filtered in the renal glomeruli.
(e) can be reincorporated into haemoglobin.

Q 1-10 Concerning bile:
(a) most of the water is absorbed from the bile in the gall bladder.
(b) a high ratio of bile salts to cholesterol favours the formation of gall stones.
(c) the cholesterol normally present in bile is derived mainly from the breakdown of steroid hormones.
(d) the volume of bile entering the duodenum is about 500 ml/24 hours.
(e) about 10% of the bile salts entering the duodenum is lost in the faeces.

Q 1-11 Bile acids:
(a) are derived from the breakdown products of haemoglobin.
(b) are secreted into the bile in the gall bladder.
(c) are concentrated in the gall bladder.
(d) are water-soluble.
(e) break down fats to fatty acids.

Q 1-12 Concerning the autonomic nerve supply to the small intestine:
(a) the parasympathetic nerve supply is through the vagus nerves.

(b) the parasympathetic nerve fibres running from the central nervous system relay at synapses before entering the gut wall.
(c) increased parasympathetic activity results in relaxation of the ileocolic sphincter.
(d) the sympathetic nerve supply leaves the central nervous system in the thoracolumbar region of the spinal cord.
(e) the sympathetic neuro-effector endings in the gut liberate acetylcholine.

Q 1-13 Concerning intestinal absorption:
(a) a volume of fluid in excess of 10 litres per day is absorbed from the gut.
(b) protein is absorbed only in the form of dipeptides.
(c) absorption of the products of protein digestion depends on active transport mechanisms.
(d) vitamin B_{12} is largely absorbed in the stomach.
(e) products of fat digestion can reach the systemic blood before reaching the liver.

Q 1-14 The ileum is the principal site for the absorption of:
(a) glucose.
(b) the products of fat digestion.
(c) bile salts.
(d) vitamin K.
(e) iron.

Q 1-15 In the colon:
(a) the secretions lack significant digestive enzymes.
(b) sympathetic stimulation results in enhanced motility.
(c) stimulation of the nervi erigentes results in increased mucous secretion.
(d) more than half the water is absorbed from the contents.
(e) bacterial synthesis of vitamin K is of vital importance.

Q 1-16 Concerning large bowel function:
(a) constipation results in the absorption of toxic substances from the bowel.
(b) defaecation is a spinal reflex.
(c) distension of the stomach with food initiates contractions of the colon.

(d) the parasympathetic nerve supply is excitatory to the musculature of the colon (excluding sphincters).
(e) the sympathetic nerve supply is excitatory to the internal (involuntary) anal sphincter.

Q 1-17 Concerning vomiting:
(a) vomiting as a component of travel sickness is primarily the result of irritation of gastro-intestinal receptors.
(b) apomorphine inhibits vomiting.
(c) the central nervous centre for the vomiting reflex is in the cerebral cortex.
(d) when the duodenum is irritated, the resultant sensation of nausea is associated with relaxation of the duodenum.
(e) the efferent pathways of the vomiting reflex involve both autonomic and somatic components.

Q 1-18 If a healthy person goes without food for a week:
(a) blood glucose is unlikely to fall below 50 mg/dl (= 2.5 mmol/l) (half normal).
(b) insulin secretion decreases.
(c) muscle metabolizes free fatty acids (FFA).
(d) the liver forms glucose from amino acids.
(e) brain metabolizes FFA.

Q 1-19 Glucose in the plasma of a normal human:
(a) remains elevated for about 2 hours after ingesting 50 g glucose.
(b) is entirely removed from all plasma flowing through the glomerular capillaries.
(c) is replenished, if its concentration decreases, from liver glycogen.
(d) is the essential energy source for the central nervous system.
(e) is bound to albumin.

Q 1-20 Functions which can be carried out ONLY in the liver include:
(a) synthesis of prothrombin.
(b) breakdown of red blood cells.
(c) conversion of glucose to glycogen.
(d) breakdown of glycogen stores to release glucose into the blood.
(e) secretion of bile salts.

Q 1-21 With reference to hepatic blood flow:
(a) about one-third of the blood flowing to the liver is arterial blood (hepatic artery).
(b) the total flow from the liver accounts for about one-quarter of the venous return to the heart (at rest and in the postabsorptive state).
(c) the flow rate decreases during exercise.
(d) portal venous and arterial blood mix in the capillaries (sinusoids).
(e) portal venous blood contains all the absorbed products of digestion.

Q 1-22 With reference to the endocrine control of carbohydrate metabolism:
(a) cortisol inhibits tissue glucose utilization.
(b) uptake of glucose into brain cells depends on an adequate amount of insulin.
(c) insulin promotes synthesis of glycogen in muscle.
(d) glucagon secretion exceeds insulin secretion during fasting.
(e) adrenaline promotes glycogenolysis.

Q 1-23 During very heavy prolonged muscular work or exercise:
(a) muscle uses up its glycogen store.
(b) blood glucose concentration falls progressively from the start.
(c) circulating catecholamines promote gluconeogenesis.
(d) arterial carbon dioxide tension (P_aCO_2) increases.
(e) plasma lactate concentration increases.

Section 2 Breathing and Gas Exchange

Q 2-01 The functional residual capacity (FRC) in the lungs of a healthy adult of average size:
(a) is about 1 litre.
(b) becomes smaller if airflow resistance increases.
(c) can be estimated using a helium dilution method.
(d) has the effect of damping fluctuations of alveolar gas concentrations during the breathing cycle.
(e) is the volume at which some airways normally begin to close during expiration.

Q 2-02 In a healthy subject, sitting upright; at rest:
(a) tidal volume is one-tenth or less of the total lung volume.

(b) the lungs inflate and deflate around a mean volume which is about one-quarter of their full capacity.
(c) if he breathes right out, small airways start to close in the lower parts of the lungs sooner than in the upper parts.
(d) if he breathes right out to residual volume (RV), the first air subsequently inhaled will enter the apical regions of the lungs.
(e) if a resistance is added, externally, to airflow, the tidal exchange will shift to a higher lung volume.

Q 2-03 An adult subject breathing air was found to have the following lung volumes:

Vital capacity	3·5 litres
Forced expiratory volume in 1 sec (FEV_1)	2.8 litres
Functional residual capacity (FRC)	1.8 litres
Residual volume (RV)	0.8 litre

(a) there is no obstruction to airflow.
(b) the subject must be abnormal.
(c) the expiratory reserve volume is 1 litre.
(d) all of these measurements could have been made using only a spirometer.
(e) there would be approximately 250 ml of oxygen in this subject's lungs at the end of a tidal expiration.

Q 2-04 The graphs represent the change in volume when an isolated lung is inflated: A-B represents a typical curve, for a lung which is normal and air-filled.

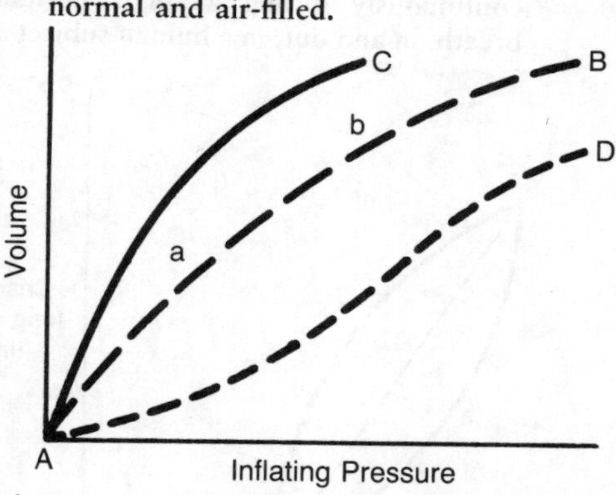

(a) the lower part of the curve, A-a shows the difficulty of initial expansion against surface tension forces.
(b) the upper part, b-B represents the volume at which the lung is most compliant.
(c) during deflation, the same pathway would be followed in reverse (B-b-a-A).

(d) the curve A-C could represent expansion of the same lung when filled with saline instead of air.
(e) the curve A-D could represent a lung with more surfactant activity.

Q 2-05 Compliance of the lungs *in vivo*
(a) depends partly on airway conductance when measured during continuous breathing.
(b) is defined as the change in volume per unit change in expanding (inflating) pressure.
(c) is greatest, for the whole lung, between residual volume and functional residual capacity.
(d) is decreased if surfactant is depleted.
(e) within the tidal range, is greater at the apex than at the base of the lungs in the upright posture.

Q 2-06 Concerning mechanical factors in breathing:
(a) there is more muscular work involved in breathing in, than in breathing out, in the tidal range.
(b) forced expiration is more difficult than forced inspiration.
(c) recoil of the chest wall assists inspiration.
(d) the 'negative pressure' holding the lungs inflated is less effective (less negative) at the bases than at the apices.
(e) respiratory muscles use about one-tenth of the whole body oxygen consumption in normal people.

Q 2-07 The diagram represents measurements made continuously during a single normal tidal breath, in and out, in a human subject at rest:

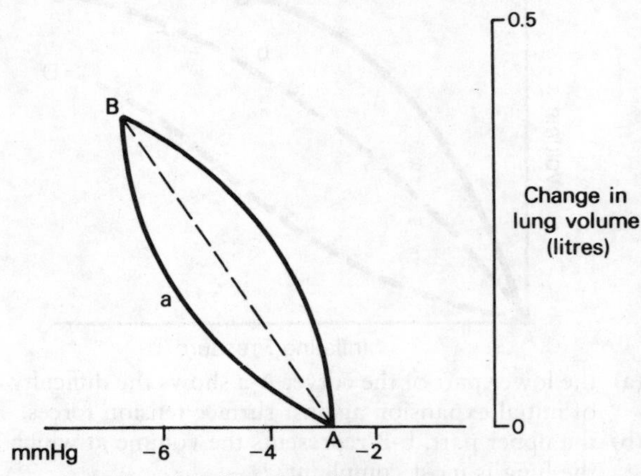

(a) the curve B-a-A represents expiration.

(b) zero on the vertical scale represents functional residual capacity.
(c) the volume change shown here could be increased by a factor of 16 in very strenuous exercise.
(d) the slope of the straight dashed line drawn from A to B measures the compliance of the lungs.
(e) the 'loop' shown would be 'fatter' if the person had increased resistance to airflow.

Q 2-08 The muscular work done during inspiration:
(a) is made less by the effect of surfactant.
(b) is greater if the elastic recoil of the lungs is greater.
(c) is greater if inspiration starts at a high lung volume, than if it is in the normal tidal range.
(d) could be lessened by bronchial dilatation.
(e) is partly spent in overcoming surface tension forces.

Q 2-09 A patient is mechanically ventilated at 8 litres/min. The anatomical dead space is estimated to be 150 ml. The stroke volume of the respirator (tidal volume) is set at 0.8 litres. The inspired gas contains 50% oxygen. CO_2 production is 195 ml/min and O_2 usage is 240 ml/min. Alveolar CO_2 is 3%:
(a) the alveolar ventilation is 6.5 litres/min.
(b) the patient is being hyperventilated.
(c) at this alveolar CO_2 concentration, the patient is likely to have cerebral vasodilatation.
(d) given that his pulmonary gas exchange is normal, his arterial oxygen tension (PO_2) should be over 40 kPa (300 mmHg).
(e) the respiratory quotient (RQ) is over 1.2.

Q 2-10 In relation to oxygen consumption in adult man and its measurement:
(a) if the subject is at rest and hyperventilates, the oxygen percentage in the expired air will decrease.
(b) during moderate muscular exercise, ventilation increases linearly as oxygen consumption increases.
(c) by the spirometer method, the oxygen consumption is measured from the rate of emptying of the spirometer bell.
(d) the average resting oxygen consumption is about one-quarter of a litre per minute.
(e) consumption of one litre of oxygen is equivalent to an energy expenditure of approximately 4.8 kcal (20 kJ).

Q 2-11 In the alveolar region of the lungs:

(a) the barrier to diffusion of gases is about 10 μm thick.
(b) surfactant-secreting cells form a continuous cytoplasmic layer lining the alveoli.
(c) alveoli are larger in the upper and smaller in the lower zones of the lungs.
(d) if any fluid escapes from capillaries, it goes into the alveoli.
(e) macrophages in the alveoli are carried to the respiratory bronchioles by the activity of cilia.

Q 2-12 Arterial hypoxaemia (decreased arterial oxygen tension) is a consequence of:

(a) hypoventilation.
(b) low haemoglobin concentration.
(c) carbon monoxide poisoning.
(d) living at high altitude.
(e) ventilation–perfusion maldistribution in the lungs.

Q 2-13 In a healthy person, the following values were found by continuous sampling of end-tidal (end-expired) gas, and can be taken to represent alveolar partial pressures:

PO_2—15 kPa (115 mmHg) PCO_2—3.5 kPa (25 mmHg)

(a) the subject was overbreathing (hyperventilating).
(b) the oxygen percentage in the inspired gas must have been higher than in room air.
(c) the arterial PCO_2 would be 3.5 kPa (25 mmHg).
(d) the arterial PO_2 would be 12 to 13.5 kPa (90 to 100 mmHg).
(e) there must be a respiratory alkalosis.

Q 2-14 A healthy person is given pure oxygen to breathe for 5 minutes:

(a) in the first minute the ventilation will be considerably depressed.
(b) in the fifth minute the ventilation will be virtually the same as it was breathing air.
(c) the arterial PCO_2 will decrease considerably.
(d) the arterial PO_2 will rise to over 80 kPa (600 mmHg).
(e) the subject would be able to hold his breath at the end of the 5 minutes for significantly longer than when he was breathing air.

Q 2-15 With reference to the control of breathing:

(a) the increase in ventilation in exercise is proportional to a rise in arterial PCO_2.

(b) the peripheral (arterial) chemoreceptors are stimulated by any form of diminished oxygen content in the arterial blood.
(c) in adult man, afferents carried in the vagus nerves play no part in the control of breathing.
(d) breathing can continue when the brain stem is the only functioning part of the brain.
(e) breathing can be influenced voluntarily via direct pathways from the cerebral cortex to spinal motoneurones.

Q 2-16 Movement of fluid out of pulmonary capillaries:
(a) is increased if surfactant is deficient.
(b) is an entirely normal phenomenon.
(c) implies movement of fluid into the alveolar spaces.
(d) is necessarily increased in rate by an increase in pulmonary capillary pressure.
(e) is necessarily increased in rate by an increase in pulmonary blood flow.

Q 2-17 Consider the following measured or estimated values for a small healthy woman. Values are shown for the subject at rest in the left column and during brisk treadmill walking in the right column:

	Rest	*Exercise*
Ventilation (minute volume)	4 l/min	30 l/min
Frequency of breathing	10/min	30/min
Mixed expired CO_2%	3.5%	?
End-expired (end-tidal) CO_2%	5%	?
Cardiac output	4.5 l/min	13.5 l/min
Arterial oxygen content	200 ml/l	200 ml/l
Oxygen consumption	190 ml/min	1.35 l/min
Heart rate	70/min	145/min

On the basis of these values and calculations which can be made from them, decide if the following are correct, in this subject at rest:
(a) the alveolar ventilation is 2.8 litres/min.
(b) the dead space/tidal volume ratio is 0.3.
(c) from calculation of the RQ it is possible to deduce that she is likely to be on a high carbohydrate diet.
(d) the mixed venous oxygen content is 142 ml/litre.
(e) she has a slight respiratory acidosis.

Q 2-18 For the same subject as in the previous question, in this subject during exercise:
(a) the tidal volume has trebled.

(b) the level of exercise is likely to represent her maximal working capacity.
(c) given that the RQ becomes 0.9 during exercise, her mixed expired gas during exercise must contain 4.05% CO_2.
(d) half the total oxygen is extracted from the blood.
(e) arterial PCO_2 is likely to be 6 to 6.6 kPa (45–50 mmHg).

Q 2-19 The values below represent measurements made on a patient, at rest, breathing air:

Blood volume 5.5 l Right atrial pressure 20 mmHg
Arterial blood pressure 100/60 mmHg
Arterial PCO_2 8 kPa (60 mmHg)
Arterial blood pH 7.34 Weight 70 kg

Are the following statements correct?

(a) He is suffering from hypovolaemic shock.
(b) There is a respiratory acidosis.
(c) There is some degree of failure of the heart as a pump.
(d) The arterial PO_2 would be lower than normal.
(e) Bicarbonate reabsorption in the kidneys is likely to be proceeding at a lower rate than normal.

Q 2-20 An individual has 13.5 g Hb/dl blood, and his Hb is 100% saturated at a PO_2 of 13.3 kPa (100 mmHg), when he is breathing air: his cardiac output is 6 l/min, and his oxygen usage is 300 ml/min (Assume Hb carries 1.34 ml O_2/g, and 0.3 ml of O_2 dissolves in 1 dl of blood at PO_2 13.3 kPa (100 mmHg)

In this person:

(a) the arteriovenous difference for oxygen is 50 ml/litre.
(b) the mixed venous oxygen content is 15 ml/dl.
(c) if he increases his oxygen usage four-fold, to 1.2 l/min, and doubles his cardiac output, the tissues must be extracting twice as much oxygen from each litre of blood.
(d) if he changes from breathing air to breathing oxygen, so that the alveolar PO_2 becomes 80 kPa (600 mmHg), the arterial oxygen content will be increased by 1.5 ml/dl.
(e) if the alveolar PO_2 becomes 80 kPa (600 mmHg), the arterial PO_2 will not rise above 16 kPa (120 mmHg).

Q 2-21 Suppose that the following values were obtained from an average-sized patient at rest, and decide whether the statements a-e are correct:

Systolic pressure
  in the pulmonary artery: 40 mmHg
  in the aorta: 115 mmHg.
Oxygen content in blood
  from the pulmonary artery: 140 ml/litre
  from the aorta: 200 ml/litre.
Whole body oxygen consumption: 0.30 litre/min.

(a) the aortic systolic blood pressure is normal.
(b) the pulmonary artery systolic pressure is normal.
(c) it can be inferred that the haemoglobin concentration is at least normal.
(d) the cardiac output is 6 litres/min.
(e) the oxygen consumption is low.

Q 2-22 Side effects of acute altitude hypoxia (or of the compensatory changes it leads to) include:

(a) increased pulmonary vascular resistance.
(b) decreased blood viscosity.
(c) hypocapnic cerebral vasoconstriction.
(d) increased bicarbonate reabsorption in the kidneys.
(e) increased heart rate.

Q 2-23 Compensations for the low partial pressure of oxygen in inspired air at high altitude include the changes listed on the left. Are the mechanisms on the right in each case appropriate?

(a) increased ventilation	carotid sinus receptor stimulation
(b) increased cardiac output	increase in sympathetic activity
(c) increased red blood cell count	increase in erythropoietin secretion
(d) shift of dissociation curve of HbO_2	increase in 2,3-diphosphoglycerate (DPG) in red cells
(e) hypocapnia	decreased tissue metabolism

Q 2-24 At 2 atmospheres ambient pressure (e.g. under water) in a healthy adult breathing air from a cylinder:

(a) the arterial oxygen tension is over twice normal.
(b) the arterial PCO_2 is about twice the normal value.
(c) the alveolar CO_2 concentration is about 2.6%.
(d) the arterial nitrogen tension is higher than normal.
(e) the volume of the lungs is about half normal.

Q 2-25 With reference to hyperbaric conditions:

(a) during a breath-hold dive the volume of the lungs decreases.

(b) diving, breathing air from a cylinder, at a depth of about 30 ft, the inspired oxygen partial pressure is about 40 kPa (300 mmHg).

(c) oxygen can be safely breathed for several hours at a depth of 100 ft.

(d) there is a danger of CO_2 narcosis, when breathing air at a pressure of 3 atmospheres.

(e) after exposure to high ambient pressure, fat people require more prolonged decompression than thin people.

Q 2-26 The graphs show the effect on alveolar PCO_2 of alterations in alveolar ventilation. The solid curve refers to a typical healthy subject at rest; his alveolar ventilation and PCO_2 are shown at a:

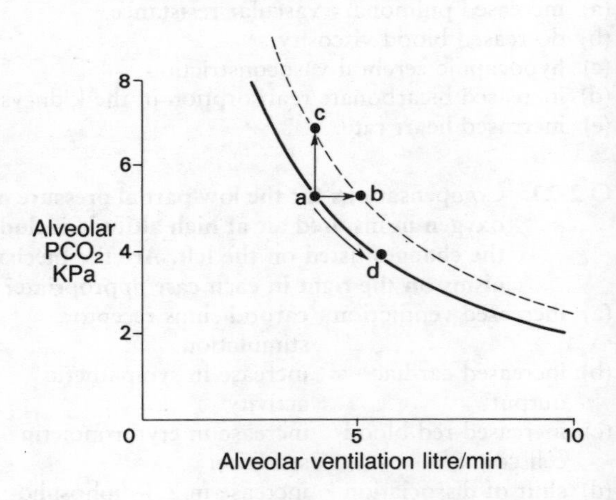

(a) the arrow a to b could represent what happens when this same subject increases his metabolic activity.

(b) the arrow a to d represents what happens during hyperventilation.

(c) the arrow a to c shows the ventilatory response to rising PCO_2.

(d) the broken curve could refer to a different subject of larger size.

(e) CO_2 production in the condition represented at (a) is a little over 200 ml/min.

Section 3 Blood, Heart and Circulation

Q 3-01 A blood sample taken from a young man showed the following values:

Packed cell volume (PCV)	**0.45 l/l**
Haemoglobin concentration	**15.2 g/dl**
Erythrocyte count	**5.0×10^{12}/l**

Are the following statements true for this subject?

(a) The red cell count is within normal limits.
(b) The mean volume of this subject's erythrocytes is 90 fl (1 litre = 10^{15} femtolitres)
(c) This blood contains about 55% plasma.
(d) The concentration of haemoglobin within the red blood corpuscles is more than twice the concentration in whole blood.
(e) If the same investigations were carried out on a patient's blood it would be possible to determine whether the patient has a macrocytic anaemia.

Q 3-02 Can the abnormality on the left be diagnosed by only the investigation on the right?

(a) Microcytosis	Red cell count
(b) Anaemia	Mean corpuscular volume (MCV)
(c) Hypochromia	Mean corpuscular haemoglobin concentration (MCHC)
(d) Leucocytosis	White cell count
(e) Erythrocytes abnormally fragile	Packed cell volume (haematocrit)

Q 3-03 The haemoglobin in blood:

(a) is completely deoxygenated in the blood drawn from an arm vein.
(b) is the major source of bile pigments.
(c) is responsible for the colour of the blood.
(d) can be estimated by adding dilute hydrochloric acid and comparing the solution with a standard.
(e) contains iron in the ferrous state.

Q 3-04 With reference to the blood:

(a) a blood film stained with Leishman's stain may be used to count the number of white cells per litre of blood.

(b) the number of platelets seen in a normal blood film is the same as the number in the original sample of blood.
(c) reticulocytes are red blood corpuscles ready for destruction by the reticuloendothelial system.
(d) the production of erythropoietin is stimulated by low oxygen tension.
(e) the life of erythrocytes in a normal human is typically 10 days.

Q 3-05 Concerning the ABO blood groups:
(a) a person of group O is a universal recipient.
(b) a person of group B usually has anti-A agglutinins in his plasma.
(c) in an incompatible blood transfusion reaction, donor cells are lysed by recipient antibodies.
(d) a severe transfusion reaction is likely to be followed by jaundice.
(e) antibodies to the A and B agglutinogens are complete cold antibodies.

Q 3-06 Concerning the rhesus (Rh) blood grouping:
(a) every baby of a Rh positive father and Rh negative mother is at risk.
(b) the second Rh positive child of a Rh negative mother is at greater risk than the first.
(c) transfusion of group A Rh positive blood for the first time to a young nullipara whose blood group is A Rh negative, is likely to cause a severe reaction.
(d) the antibodies which cross the placenta from maternal to fetal circulation are incomplete antibodies.
(e) an appropriate treatment for a neonate suffering from erythroblastosis neonatorum is transfusion with maternal blood.

Q 3-07 As blood passes through systemic capillaries:
(a) its pH rises.
(b) bicarbonate ions pass from the red cells to the plasma.
(c) the concentration of chloride ions in the red cells falls.
(d) its oxygen dissociation curve shifts to the right.
(e) the velocity of blood flow is less than in the aorta.

Q 3-08 With reference to the normal human heart:
(a) the most quickly conducting fibres in the heart are the Purkinje fibres.

(b) the last part of the ventricle to be activated is the apex.
(c) the duration of the action potential in a ventricular muscle fibre is about the same as in a skeletal muscle fibre.
(d) the T-wave of the ECG occurs at the beginning of the absolute refractory period of the ventricle.
(e) left axis deviation leads to an abnormally large R wave in Lead I (left arm—right arm).

Q 3-09 The 3 traces show records from the standard leads of an electrocardiogram (ECG). In these traces:
(a) period a corresponds to inspiration.
(b) when the heart rate changes during the record (lead II), the P–T interval changes less than the T–P interval.
(c) the record could have been taken from a patient with a denervated heart.
(d) the conduction time between the sino-atrial and the atrioventricular nodes is greater than normal.
(e) the direction of the electrical vector at the time of ventricular depolarization is approximately the same as that at the time of ventricular repolarization.

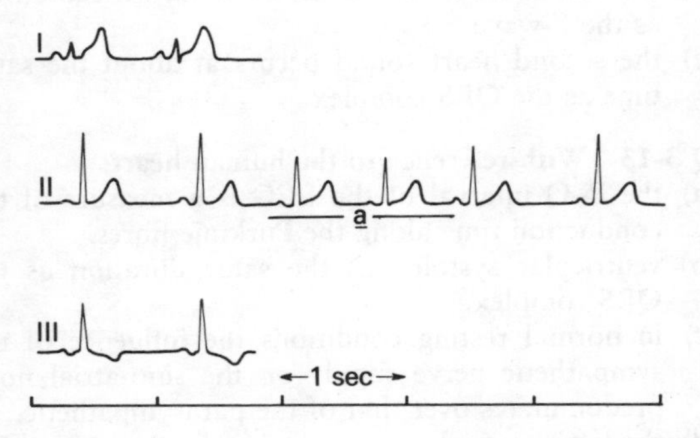

Q 3-10 Concerning the electrocardiogram (ECG) of an adult human:
(a) the P wave coincides with depolarization of the atria.
(b) the Q wave coincides with repolarization of the atria.
(c) a P–R interval of 0.3 seconds indicates impaired conduction.
(d) the R wave coincides with depolarization of the apex of the heart

(e) during the isopotential (isoelectric) phase between the S and T waves, the intracellular potential in ventricular muscle cells is positive with respect to the interstitial fluid.

Q 3-11 With reference to the electrical events in the cardiac cycle:

(a) the ventricular action potential lasts about 0.3 seconds at rest.
(b) high extracellular potassium may result in ventricular fibrillation.
(c) the QRST complex of the ECG has the same wave form as that which would be recorded from a microelectrode inside a ventricular muscle fibre.
(d) vagal stimulation reduces the rate of depolarization in pacemaker cells during diastole.
(e) the upstroke of the ventricular action potential is due to an increase in potassium permeability of the muscle membrane.

Q 3-12 With reference to the ECG:

(a) the QRS complex is produced by depolarization of the ventricles.
(b) the Q–T interval gives an approximate indication of the duration of ventricular systole.
(c) the aortic valve is closed at the time of the P wave.
(d) the first heart sound occurs at about the same time as the P wave.
(e) the second heart sound occurs at about the same time as the QRS complex.

Q 3-13 With reference to the human heart:

(a) the P–Q interval of the ECG is a measure of the conduction time along the Purkinje fibres.
(b) ventricular systole has the same duration as the QRS complex.
(c) in normal resting conditions the influence of the sympathetic nerve supply on the sino-atrial node predominates over that of the parasympathetic.
(d) the left ventricular pressure rises by 8 to 10 mmHg during the phase of isometric contraction.
(e) the rate of emptying of the left ventricle is fairly uniform throughout the ejection phase of ventricular systole.

Q 3-14 In the adult human electrocardiogram represented here:

(a) judging by the interval between T and P relative to the duration of the whole cycle, the pulse rate is likely to be nearer 60 than 120 beats per minute.

(b) at the time corresponding to arrow a, the rate of filling of the ventricles would be at its peak.
(c) the arrow b indicates a time at which there is a rise in potassium permeability in ventricular muscle cells.
(d) the line c represents the period during which the atria are contracting.
(e) the line d represents the duration of ventricular systole.

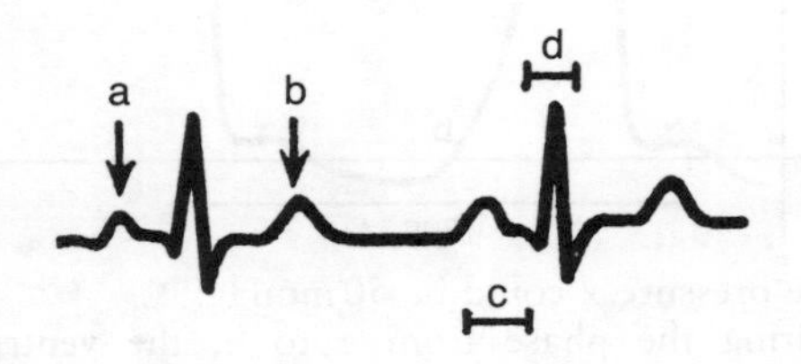

Q 3-15 Concerning the cardiac cycle of a normal young healthy adult human who is reclining and breathing quietly:
(a) the peak pressure in the right ventricle is about 25 mmHg.
(b) the lowest pressure in the right ventricle is about 10 mmHg.
(c) the mean jugular venous pressure is about 10 mmHg.
(d) the peak of the v-wave of the a c v complex occurs immediately before the atrioventricular valve opens.
(e) the first heart sound occurs early in atrial systole.

Q 3-16 With reference to the mechanical events in the cardiac cycle in a normal adult human:
(a) the left ventricle ejects more blood per beat than the right ventricle.
(b) the mitral valve opens when the left atrial pressure exceeds the left ventricular pressure.
(c) during strenuous work in a healthy subject, the left ventricle may contain, at the end of diastole, twice as much blood as it does at rest.
(d) the pulmonary valve opens when the right ventricular pressure reaches 20 to 25 mmHg.
(e) during diastole, the left ventricular pressure is about 70 mmHg.

Q 3-17 The graph represents a record of pressure changes in the right ventricle in a normal resting adult human:

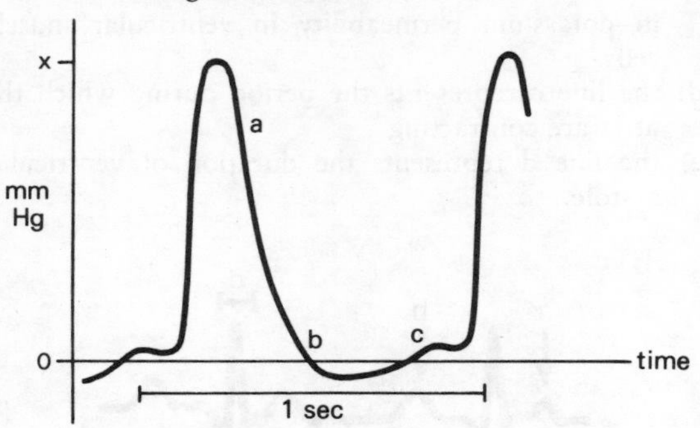

(a) the pressure x could be 50 mmHg.
(b) during the phase from a to b, the ventricle is relaxing isometrically.
(c) near b, the pulmonary valve would open.
(d) during the phase from b to c, the atrioventricular valve would be open.
(e) the peak of the pressure curve is simultaneous with the R deflection of the ECG.

Q 3-18 The diagram shows pressure-volume relations in the left ventricle of a normal human at rest. The loop 1–2–3–4 represents the sequence of changes in pressure and volume during one cardiac cycle:

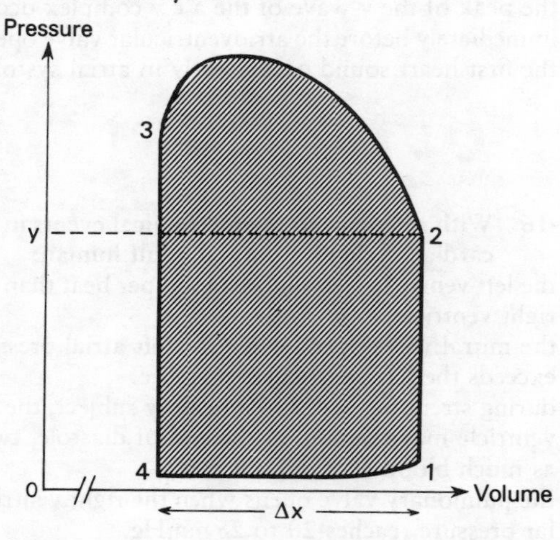

(a) the curve 2–3 represents the phase of isovolumetric contraction of the ventricles.

(b) the duration of phase 4–1 would be greater than the total duration of the rest of the loop.
(c) the volume change from 4 to 1 is about 10 ml.
(d) the pressure y at point 2 could be 70 mmHg.
(e) the shaded area represents the cardiac output.

Q 3-19 The curve AaB represents the peak systolic pressures which would be developed if the isolated left ventricle were to contract isovolumetrically at successively greater initial volumes. AbB shows the corresponding pressures and volumes during diastole. The loop 1–2–3–4 represents on the same scale one normal cardiac cycle *in vivo* in a healthy adult human:

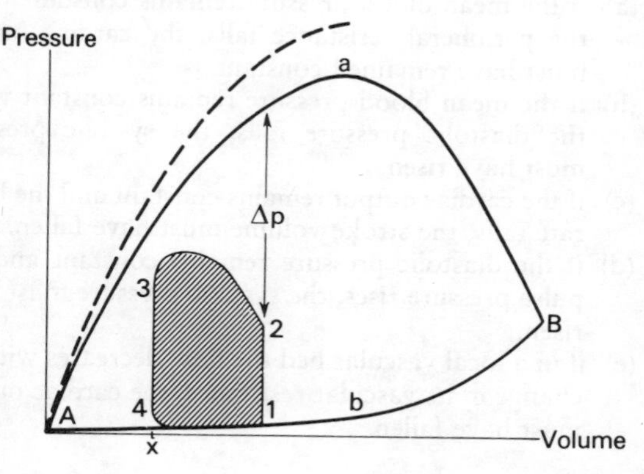

(a) the point x represents the end-diastolic volume.
(b) exercise would shift the line 3–4 to the left.
(c) the rate of ejection would be lower if Δp were smaller.
(d) sympathetic activity would shift the curve Aa to the dashed curve.
(e) administration of atropine would shift curve Aa to the dashed curve.

Q 3-20 With respect to the human heart:
(a) the spread of excitation through the wall of the ventricles is from the endocardial surface outwards.
(b) vagal stimulation decreases the force of ventricular contraction.
(c) sympathetic stimulation increases the force of atrial contraction.
(d) in exercise, systole shortens more than diastole.
(e) in the resting subject, denervation of the heart would result in a rise in heart rate.

Q 3-21 Concerning dynamics of the circulation:

(a) the resistance of a rigid cylindrical tube is inversely proportional to the square of its diameter.
(b) the arterioles contribute most of the resistance of the vascular bed.
(c) the pulse pressure measured in a medium sized artery is less than that in the aorta.
(d) if the elastic tissue in the large arteries is replaced by inelastic tissues, the pulse pressure rises.
(e) if the mean arterial blood pressure remains constant and the arterioles supplying a small vascular bed constrict, then the pressure in the capillaries supplied by those arterioles rises.

Q 3-22 Concerning flow, resistance and pressure:

(a) if the mean blood pressure remains constant when the peripheral resistance falls, the cardiac output must have remained constant.
(b) if the mean blood pressure remains constant when the diastolic pressure falls, the systolic pressure must have risen.
(c) if the cardiac output remains constant and the heart rate falls, the stroke volume must have fallen.
(d) if the diastolic pressure remains constant and the pulse pressure rises, the systolic pressure must have risen.
(e) if in a local vascular bed the flow decreases with no change in its vascular resistance, the cardiac output must have fallen.

Q 3-23 The graph represents a record of intra-arterial pressure from a healthy individual:

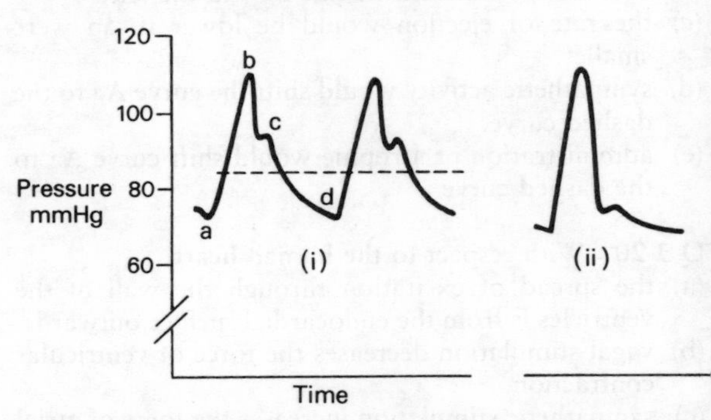

(a) the dotted line is a good approximation to the mean arterial pressure.
(b) the rate of change of pressure from a to b is related to the force of contraction of the left ventricle.

(c) during the period b to c, the left ventricle would be relaxing isovolumetrically.
(d) the rate of change of pressure from c to d depends only on the peripheral resistance.
(e) record (ii) is from a more peripheral artery than record (i).

Q 3-24 With reference to the regulation of arterial blood pressure and cardiac output:
(a) if cardiac output decreases, the arterial blood pressure will necessarily be decreased also.
(b) during muscular exercise, the total peripheral resistance is unchanged.
(c) increase in the force of ventricular contraction is one effect of an increase in sympathetic activity.
(d) the main factor causing the increase in blood flow to exercising muscles is an increase in arterial blood pressure.
(e) blood flow to the heart muscle (coronary flow) varies with the cardiac output.

Q 3-25 With reference to the regulation of arterial blood pressure and cardiac output:
(a) mean arterial pressure rises significantly during moderately severe sustained isometric exercise.
(b) tilting a normal person from the horizontal to the vertical position results in an increase in heart rate.
(c) constriction of veins is one of the responses to decreased arterial baroreceptor stimulation.
(d) when the arterial blood pressure falls, ventilation decreases.
(e) a decrease in ventricular end-systolic volume is one effect of an increase in sympathetic activity.

Q 3-26 When a normal person lies down:
(a) heart rate settles to a higher level than when standing.
(b) venous return is immediately increased.
(c) cerebral blood flow settles to a higher level than when standing.
(d) blood flow in the apices of the lungs increases.
(e) lower limb veins constrict actively.

Q 3-27 With respect to blood flow in different regions:
(a) blood flow to the brain varies with the arterial blood pressure.
(b) blood flow to the myocardium is greater during diastole than during systole.

(c) blood flow to the kidneys is high relative to their oxygen requirement.
(d) voluntary hyperventilation increases the blood flow to the brain.
(e) the increase of skeletal muscle blood flow during exercise is mainly due to decreased sympathetic noradrenergic vasoconstriction.

Q 3-28 Following moderate blood loss (e.g. 500 ml):
(a) the arterial blood pressure may be normal.
(b) skin vessels are constricted.
(c) lymph flow from peripheral tissues is diminished.
(d) the maintenance of brain blood flow is mediated by cerebral vasoconstriction.
(e) the hypothalamic osmoreceptors are stimulated.

Q 3-29 A normal 50-kg woman loses 1 litre of blood:
(a) her blood volume is depleted by about 35%.
(b) the loss of red blood cells would amount to approximately 5×10^{12}.
(c) heart rate would be likely to be over 100 per minute.
(d) immediately after the haemorrhage, the haematocrit value would be increased.
(e) the absorption of iron from the intestinal tract would subsequently increase.

Q 3-30 With respect to different types of blood vessels:
(a) veins are more compliant than arteries of similar size (i.e. distend more for a given increase in intravascular pressure in the physiological range).
(b) the wall of a capillary consists of endothelium and basement membrane only.
(c) capillaries in all regions of the body show fenestrations in their walls.
(d) all arterial vessels have smooth muscle in their walls.
(e) arteries and arterioles are the only blood vessels innervated by sympathetic nerve fibres.

Q 3-31 With reference to capillaries:
(a) fenestrations are present in the renal glomerular capillaries.
(b) cerebral capillaries have microscopically visible 'leaks'.
(c) when skeletal muscle is active, there are more capillaries open than when it is at rest.

(d) the diameter of a capillary in skeletal muscle is usually about twice the diameter of a red blood cell.
(e) capillaries in the dermis actively constrict in response to sympathetic stimulation.

3-32 During any rhythmic exercise (e.g. running):
(a) the maximal heart rate which can be reached decreases with fitness.
(b) training increases the amount of work which can be done at a given heart rate.
(c) the cardiac output can increase by a factor of 15.
(d) pulmonary vascular resistance increases.
(e) the diastolic blood pressure increases considerably.

Q 3-33 With reference to adjustments in exercise:
(a) an increase in muscle blood flow begins after the first half minute of exercise.
(b) cerebral blood flow increases if the exercise causes systolic arterial blood pressure to rise.
(c) body temperature may rise measurably.
(d) lymph flow from the exercising muscles increases.
(e) visceral blood flow decreases.

Q 3-34 During intra-uterine life:
(a) all the fetal blood returning from the placenta flows directly into the inferior vena cava.
(b) fetal blood carries more oxygen than maternal blood at a low PO_2.
(c) many antibodies are transferred freely between maternal and fetal circulations.
(d) blood in the right side of the fetal heart is slightly better oxygenated than blood in the left side.
(e) pulmonary vascular resistance is higher than after birth.

Section 4 Cell Structure and Function

Q 4-01 With reference to microscopic structure:
(a) the blood/air barrier in lung alveoli is typically less than 1 μm thick.
(b) the trachea is lined by ciliated epithelium.
(c) there are mucus-secreting cells along the whole length of the alimentary tract.
(d) cytoplasmic bridges connect the cells of single-unit smooth muscle.

(e) neutrophil polymorphs are usually the commonest of the white cells in circulating blood.

Q 4-02 With reference to microscopic structure and function:

(a) the T-tubes of skeletal muscle release Ca^{++} into the muscle fibre.
(b) the oxyntic cells of the gastric mucosa secrete pepsinogen.
(c) the type II alveolar cells of the lung are responsible for the secretion of surfactant.
(d) the interstitial cells of the testis have an endocrine function.
(e) the myoepithelial cells of the mammary gland are concerned with lactogenesis (milk formation).

Q 4-03 With reference to the functional significance of microscopic structure:

(a) the function of microvilli of small intestinal mucosal cells is entirely to provide a large surface for absorption of the products of digestion.
(b) the microvilli (brush border) of the proximal convoluted tubule in the kidney are associated with passive movement of Na^+ into the cells down a diffusion gradient.
(c) cells of the juxtaglomerular apparatus secrete renin.
(d) microvilli on the surface of thyroid follicular cells aid in the reabsorption of colloid from the store.
(e) in the stomach, intrinsic factor is secreted by cells in the mucosa of the antrum (the distal fifth).

Q 4-04 With reference to mitochondria:

(a) they are more numerous in cardiac than in any other kind of muscle.
(b) they are one type of intracellular vesicle.
(c) they are most numerous in cells whose primary source of energy is anaerobic glycolysis.
(d) they are numerous in red blood cells.
(e) they contain chromosomes.

Q 4-05 Active transport across cell membranes:

(a) is increased by hypothermia.
(b) transfers hydrogen ions into gastric juice against a concentration gradient.
(c) requires energy production by the cell.
(d) aids hydrogen ion secretion by kidney tubule cells.
(e) prevents an excess of water from entering the cell.

Q 4-06 The equilibrium potential across a cell membrane for an ionic species:
(a) is the membrane potential at which there is no net movement of these ions across the membrane, assuming the ion to be permeant.
(b) is the voltage that the membrane potential would assume if the membrane were permeable to that ion and to no others.
(c) would be zero if the concentration of the ion on one side of the membrane were ten times its concentration on the other side.
(d) is necessarily zero if the ion is impermeant.
(e) has approximately the same value for sodium and potassium ions in resting nerve fibres.

Q 4-07 With respect to the ways in which substances enter and leave cells:
(a) potassium ions move in and out through water-filled 'pores'.
(b) glucose enters by dissolving in the cell membrane.
(c) amino acids enter by a carrier-mediated transport mechanism.
(d) all healthy cells have sodium pumps which extrude sodium ions against a concentration gradient.
(e) only water-soluble substances can cross cell walls.

Q 4-08 Concerning osmotic pressure:
(a) A solution of 5 g/litre of haemoglobin (MW 67 000) has a higher osmotic pressure than a solution of 5 g/litre of urea (MW 60).
(b) The total osmotic pressure of the plasma is about 50 mmHg.

If erythrocytes are placed in a hypertonic solution of sodium chloride:
(c) the cells swell.
(d) haemolysis may occur.
(e) water diffuses out of the cells.

Section 5 Nervous System and Muscle

Q 5-01 Concerning cutaneous sensation in the human:
(a) a human can sense cold in skin in which the only sensory nerve endings are naked nerve endings.
(b) the distribution of touch sensitivity is punctate.
(c) skin receptors relay to the spinal cord via axons which conduct at about 100 m/sec.

(d) two-point discrimination is better developed in the skin of the forefinger than in that of the arm.
(e) Pacinian corpuscles lie in the epidermis.

Q 5-02 Are these associations appropriate between receptor and stimulus?

(a) Muscle spindles	Tapping the patellar tendon
(b) Receptor cells in the utricle	Linear acceleration
(c) Receptor cells in the auditory system	Movement of the basilar membrane of the cochlea
(d) Rod photoreceptor	Red light
(e) Pacinian corpuscle	Heat

Q 5-03 With reference to the muscle spindle:
(a) the primary sensory ending is stimulated by shortening of the extrafusal fibres around the muscle spindle.
(b) the primary sensory ending in a given muscle is stimulated when an antagonist muscle shortens.
(c) the primary sensory ending is stimulated by contraction of the intrafusal muscle fibres.
(d) the contractile component of the spindle is smooth muscle.
(e) the motoneurones supplying the intrafusal muscle fibres are usually smaller than those supplying the extrafusal fibres.

Q 5-04 The figure shows the transmembranal potential for a stretch receptor before, during and after a stretch. Are the following correct?

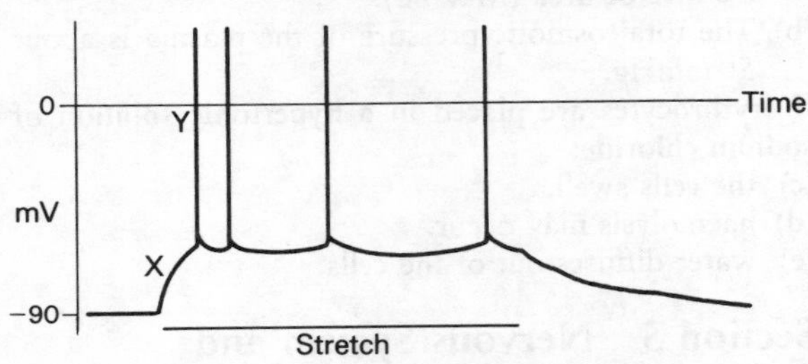

(a) The initial response to stretch (at X) is a generator potential.
(b) The initial response is a depolarization.
(c) Stretch produces at Y a high amplitude generator potential.
(d) Stretch produces a train of action potentials which demonstrates adaptation.
(e) The peak voltage of the action potentials depends on the duration of the stimulus.

Q 5-05 Concerning human vision:

(a) a light stimulus which is close to the threshold for dark-adapted eyes is more likely to be seen if it is slightly off the visual axis than if it is exactly on the visual axis.
(b) the normal dark-adapted subject can recognize the colour of near-threshold light stimuli.
(c) in a well-illuminated room, an emmetrope with one dioptre of accommodation looks with one eye at a page of writing held at 30 cm. His facility of reading will improve if he looks through a disc with a pinhole in the middle.
(d) the direct light reflex depends on the integrity of the visual area of the cerebral cortex.
(e) if a normal subject changes his foçus point from 25 cm to infinity, there is an accompanying constriction of the pupils.

Q 5-06 Concerning vision and eye movements:

(a) a lesion of the 6th cranial nerve results in inward deviation of the ipselateral eyeball.
(b) when a subject reads, his eyes move smoothly to scan the page.
(c) if a subject watches an object moving at a constant velocity, his eyes move smoothly to follow the object.
(d) the lateral semicircular canals are more closely connected with the eye muscles causing lateral eye movements than with those causing vertical eye movements.
(e) nystagmus can be induced in a normal subject without stimulation of the semicircular canals.

Q 5-07 Concerning normal human vision:

(a) for most normal humans, visual input is necessary for accurate alignment of the visual axis.
(b) during dark-adaptation, the sensitivity of the eye increases by about tenfold.
(c) dark adaptation depends principally on pupillary dilatation.
(d) the lens contributes more than the anterior surface of the cornea to the refractory power of the eyeball.
(e) if the eyes are open under water, they can form focussed images of an illuminated underwater object on the retinae.

Q 5-08 The diagram shows pathways concerned with vision and pupillary reflexes:

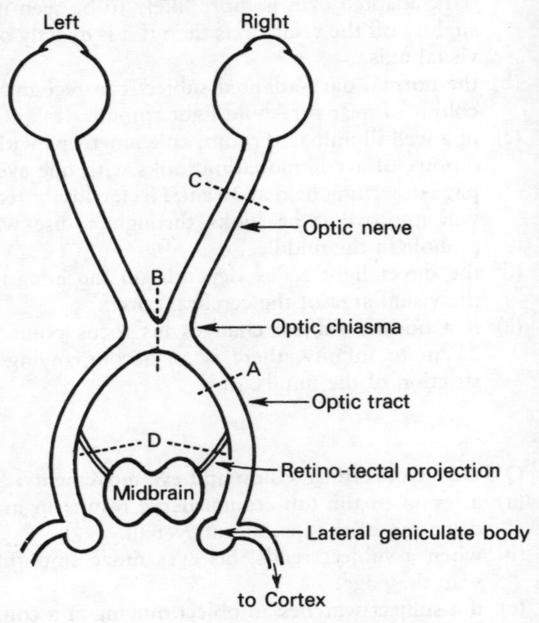

(a) a lesion at A will cause a left homonymous hemianopia.
(b) a lesion at B will cause a bitemporal heteronymous hemianopia.
(c) with a lesion at C, light shone into the left eye will cause pupillary constriction in the right eye (consensual light reflex).
(d) lesions at D will abolish the direct light reflexes on both sides.
(e) lesions at D will abolish the reflex pupillary constriction with accommodation (the accommodation-convergence reflex) on both sides.

Q 5-09 Concerning the ciliary muscles and accommodation of the eye:

(a) the increase in power of the eye needed for near vision is achieved mainly by an increase in curvature of the posterior surface of the lens.
(b) contraction of the ciliary muscle causes the lens to get thinner.
(c) ciliary muscle is multiple unit smooth muscle.
(d) circulating adrenaline causes the ciliary muscle to contract.

(e) the nerve supply to the ciliary muscle runs with the optic nerve.

Q 5-10 An individual has 1 dioptre (1D) of hypermetropia (hyperopia) and his range of accommodation is 2 dioptres: (1D = 1/focal length in metres)

(a) his near point is 1 m in front of his cornea.
(b) his far point is 1 m behind his cornea.
(c) he can clearly see distant objects.
(d) if his accommodation is paralysed by atropine applied to the eye, he can clearly see objects 1 m in front of him.
(e) if his accommodation is paralysed, a +3D spectacle lens will allow him to see clearly an object 50 cm in front of him.

Q 5-11 Consider an emmetropic individual whose range of accommodation is 2 dioptres.

(a) his near point is 1 m in front of his cornea.
(b) his far point is 1 m behind his cornea.
(c) he can clearly see distant objects.
(d) if his accommodation is paralysed by atropine applied to the eye, he can clearly see objects 1 m in front of him.
(e) if his accommodation is paralysed, a +2D spectacle lens will allow him to see clearly an object 50 cm in front of him.

Q 5-12 Concerning the nerve supply to the intrinsic eye muscles:

(a) the dilator pupillae is supplied by the sympathetic nervous system.
(b) the sympathetic nerve supply to the eye leaves the central nervous system at the level of the thoracic spinal cord.
(c) the sympathetic fibres run from the spinal cord to the eye muscles without synaptic relay.
(d) the parasympathetic nerve supply to the eye leaves the central nervous system at the level of the medulla oblongata.
(e) the cholinergic receptors at the neuro-effector junction between the parasympathetic nerve fibres and smooth muscle are of the nicotinic type.

Q 5-13 Concerning nystagmus:

(a) the visual image is suppressed during the fast phase of nystagmus.
(b) the fast phase is a ramp movement.
(c) the fast phase of nystagmus depends on the integrity of the cerebral cortex.

(d) the fast phase of nystagmus is in a direction such as to stabilize the visual image on the retina.
(e) nystagmus only occurs if the eyes are open.

Q 5-14 Consider a normal subject sitting in a Barany chair, with head held upright and stationary with respect to the body:

(a) when the chair is rotated, the semicircular canals most involved in sensing this rotation are the lateral ones.
(b) at the start of a rotation at a constant angular velocity, the slow component of the nystagmus which the subject exhibits will be in the opposite direction from that of the rotation.
(c) if the subject closes his eyes and the rotation continues at a constant angular velocity for half a minute, the subject will continue to be aware of the rotation.
(d) if the chair is abruptly stopped, there is a nystagmus with slow component in the same direction as the previous rotation.
(e) when rotation stops, and the subject stands, he tends to fall forward.

Q 5-15 If cold water is introduced into the external auditory meatus of a subject lying supine to elicit vestibulo-ocular reflexes:

(a) reflex eye movements will be generated in a normal subject within 2 seconds.
(b) the reflex eye movements produced in a normal subject are due to direct transmission of mechanical forces from the external meatus to the semicircular canals.
(c) the slow phase of nystagmus in a normal subject will be towards the side which is cooled.
(d) a slow deviation of the visual axis occurring without the fast phase indicates midbrain damage.
(e) caloric testing of this type can result in eye movements, even in unconscious patients.

Q 5-16 A human stands on the deck of a rolling boat. Receptors contributing to the postural adjustments necessary for the maintenance of an upright posture include:

(a) pressure receptors in the skin of the soles of the feet.
(b) proprioceptors in the muscles of the legs.
(c) receptors in the capsules of the joints between the upper cervical vertebrae.
(d) hair cells in the otolith organs of the vestibules.
(e) hair cells in the semicircular canals.

Q 5-17

(a) If a vibrating tuning fork is placed firmly on the skin over the mastoid process of a normal subject and its vibration is allowed to decrease until it just ceases to be audible, the subject will still be able to hear the note when the fork is held to his ear.
(b) The frequency of the highest note which a normal person can hear decreases as he gets older.
(c) The differential sensitivity to a change in pitch at a given intensity above threshold is linearly related to frequency.
(d) The otolith organs are sensitive to the orientation of the head in space.
(e) The receptors in the semicircular canals are sensitive to gravitational forces.

Q 5-18 With reference to hearing in a normal subject:

(a) contraction of the tensor tympani muscle decreases sound transmission through the middle ear.
(b) if two tones of equal amplitude, each with a frequency of 200 Hz but differing in phase by 1 msec are heard by the two ears, the subject can detect this difference.
(c) the intensity threshold for hearing is relatively independent of the frequency of a sound, in the range 50 Hz to 12 kHz.
(d) the just perceptible difference in intensity of a 1 kHz tone is relatively independent of the intensity level at which it is tested.
(e) the percentage change in frequency of a sine wave sound which is just detectable is less than the percentage change in amplitude of the sound.

Q 5-19 With respect to the ear and hearing:

(a) the lever system of the ossicles of the middle ear serves to amplify the movement of the ear drum.
(b) high pitched sounds are detected by receptors at the apical end of the cochlea.
(c) the foot-plate of the stapes fits into the oval window.
(d) contraction of the stapedius muscle causes an increase in the apparent intensity of a sound.
(e) a subject in whom the middle ear ossicles become fixed will still be able to hear in the same ear the sound from a vibrating tuning fork held against his mastoid promontory.

Q 5-20 If the neck is flexed whilst the trunk remains stationary:

(a) the pattern of activity in vestibular receptors remains the same.
(b) the resultant vestibular reflexes would include an increase in extensor tonus in the hindlimbs.
(c) the resultant neck reflexes would include an increase in extensor tonus in the hindlimbs.
(d) together, the vestibular and neck reflexes would produce a greater change in extensor tonus in the hindlimbs than either of the reflexes alone.
(e) the tonus in the superior rectus muscles in the orbits would increase.

Q 5-21 The action potential in nerve:

(a) is initiated by a depolarization of the membrane.
(b) is a change in membrane potential towards the equilibrium potential for sodium ions.
(c) involves a decrease in the membrane permeability to potassium ions.
(d) is associated with an increase in the electrical resistance of the membrane.
(e) is propagated along the axon by means of the release of acetylcholine.

Q 5-22 Concerning the nerve fibres of higher mammals:

(a) the resting potential is within 20 mV of the potassium equilibrium potential.
(b) the upstroke of the action potential is largely due to influx of sodium ions.
(c) the duration of the upstroke of the action potential is 8 to 10 msec.
(d) the peak of the action potential is within 20 mV of the chloride equilibrium potential.
(e) the downstroke of the action potential is largely due to increased activity of the sodium pump.

Q 5-23 With reference to conduction along nerve axons:

(a) conduction velocity in a particular axon is related to the strength of the stimulus.
(b) in myelinated axons the action potential is generated only at the nodes of Ranvier.
(c) if a stimulus is applied half way along a motor axon, the action potential will travel only in the direction towards the muscle it supplies.

(d) conduction velocity in somatic motor nerve fibres is in the range 1 to 10 m/sec.
(e) the smaller the diameter of the axon, the greater is the conduction velocity.

Q 5-24 With reference to the medial popliteal nerve in a normal human:
(a) the largest nerve fibres have a diameter of about 20 μm.
(b) the largest nerve fibres are efferent fibres.
When an electrical stimulus is applied percutaneously to the nerve:
(c) with a threshold stimulus, the smaller nerve fibres will be preferentially stimulated.
(d) the M response recorded from electrodes over the gastrocnemius muscle is generated by action potentials in the extrafusal muscle fibres.
(e) the H response recorded from electrodes over the gastrocnemius muscle is generated by action potentials in the intrafusal muscle fibres.

Q 5-25 Immediately after complete division of a mixed peripheral nerve:
(a) the denervated muscles exhibit the characteristic features of an 'upper motoneurone lesion'.
(b) there is loss of sensation in the denervated area of skin.
(c) the denervated area of skin will be cooler than the surrounding area.
(d) the sweat glands in the denervated skin will respond to an increase in temperature in the hypothalamus.
(e) the cut nerve fibres of the central stump are capable of regenerating along the nerve sheath.

Q 5-26 Concerning skeletal muscle in an adult human:
(a) when a fibre shortens, the length of the I band decreases.
(b) when a fibre shortens, its diameter increases.
(c) *in vivo*, the maximum shortening is by about 30%.
(d) the force of contraction is directly proportional to the rate of shortening.
(e) each fibre is supplied by branches of several motor nerve fibres.

Q 5-27 Are the following common to all three types of muscle—skeletal, cardiac and smooth?
(a) They undergo summation and tetanus when rapidly stimulated.

(b) They show striations under the light microscope.
(c) Action potentials are generated when their membranes are depolarized.
(d) The interaction of actin, myosin and ATP is responsible for generation of force by the muscle cells.
(e) An increase in free Ca^{++} concentration inside the cell is necessary for contraction.

Q 5-28 With reference to motor units in skeletal extrafusal muscle in an adult mammal:
(a) from the point of view of neural control, a motor unit is the functional unit in a muscle.
(b) the number of muscle fibres in a motor unit is about the same in all muscles.
(c) one way in which the strength of contraction of a muscle is increased is by recruitment of more motor units.
(d) steady tension during voluntary movements results from the fact that active motor units always produce fused tetanic contractions.
(e) an action potential in a nerve axon will normally excite every muscle fibre making up that motor unit.

Q 5-29 Concerning single-unit smooth muscle:
(a) it may show spontaneous rhythmic activity.
(b) it can contract to one-tenth of its original length and continue to exert tension.
(c) it may be stimulated to contract by chemicals circulating in the blood.
(d) it may be stimulated to contract by neural activity.
(e) its contraction may be inhibited by neural activity.

Q 5-30 Concerning neuromuscular transmission in man:
(a) calcium entry into the presynaptic terminal is an essential step in the release of transmitter following an action potential in the motor nerve.
(b) succinylcholine causes a depolarization of the postsynaptic membrane.
(c) blockade of transmission by curare is by competitive inhibition of the receptors on the postsynaptic membrane.
(d) the blockade of transmission by curare can be antagonized by an anticholinesterase.
(e) spontaneous release of transmitter occurs in the absence of action potentials in the motor nerve.

Q 5-31 An excitatory postsynaptic potential (EPSP):

(a) consists of a change of the membrane potential towards the equilibrium potential for potassium ions.
(b) in one neurone may occur at about the same time as IPSPs in other neurones influenced by the same afferent nerve.
(c) may summate with other postsynaptic potentials generated on the membrane of the same nerve cell.
(d) has the same ionic mechanism as the action potential in nerve.
(e) is an essential step in the activation of motoneurones during a stretch reflex.

Q 5-32 Inhibitory postsynaptic potentials (IPSPs) generated on a motoneurone by activity in Renshaw cells:

(a) are mediated by acetylcholine as transmitter.
(b) consist of a hyperpolarization of the postsynaptic membrane.
(c) are directly produced by increased activity of an ionic pump.
(d) occur in a motoneurone only as a result of that motoneurone firing an action potential.
(e) involve the same synapses as those activated by the orthodromic inhibitory pathway.

Q 5-33 Inhibitory postsynaptic potentials (IPSPs) generated on a motoneurone by activation of the orthodromic inhibitory pathway:

(a) are probably mediated by glutamate as transmitter.
(b) consist of a change in membrane potential towards the equilibrium potential for sodium ions.
(c) may be generated by synaptic terminals of primary afferent nerve fibres.
(d) may be recorded from an electrode in the dorsal root ganglion.
(e) occur as part of the response to passive stretching of the muscle supplied by that motoneurone.

Q 5-34 Presynaptic inhibition:

(a) can result from depolarization of the terminals of the first-order afferent nerve fibres.
(b) is recorded as a change in membrane potential in motoneurones.
(c) is due to a reduction in the amount of transmitter released by the presynaptic terminals.

(d) is due to a block of conduction of the action potential in the presynaptic terminals.
(e) has the same ionic mechanism as IPSPs.

Q 5-35 With reference to junctional transmission:
(a) at the motor end plate between motor nerve and muscle fibre, the transmitter is acetylcholine.
(b) at the synapse between an afferent nerve and a motoneurone in the anterior horn of the spinal cord, the excitatory transmitter is acetylcholine.
(c) a selective increase in chloride permeability of the motoneuronal membrane will result in increased excitability of the neurone (as assessed by the effectiveness of excitatory synaptic drive).
(d) strychnine causes convulsions by blocking the effects of inhibitory synaptic transmitters.
(e) calcium entry into presynaptic terminals is an essential step in the release of neurotransmitters.

Q 5-36 A human somatic lower motoneurone in the spinal cord:
(a) usually innervates only one muscle fibre.
(b) lies on the same side of the body as the muscle which it innervates.
(c) lies on the same side of the body as spindle receptors which cause its monosynaptic activation.
(d) has synapses on its cell body.
(e) has synapses on its dendrites.

Q 5-37 Concerning the knee-jerk reflex in a normal adult human:
(a) the reflex response time is about 1 msec.
(b) the reflex response time is about the same as the subject's reaction time.
(c) the reflex is monosynaptic.
(d) the associated inhibition of motoneurones supplying antagonist muscles is monosynaptic.
(e) the amplitude of the reflex contraction can be voluntarily influenced by the subject.

Q 5-38 The dorsal (or posterior) column:
(a) consists largely of second-order neurones.
(b) carries most of the central afferent fibres which subserve temperature sensation.
(c) carries fibres mainly from receptors on the ipselateral side of the body.

(d) carries fibres many of which terminate on cells in the gracile and cuneate nuclei.
(e) is one of the main input pathways to the cerebellum.

Q 5-39 The lateral spinothalamic tract:
(a) consists largely of second-order neurones.
(b) carries most of the central afferent fibres which subserve temperature sensation.
(c) carries fibres mainly from receptors on the ipselateral side of the body.
(d) carries fibres most of which terminate on cells in the thalamus.
(e) is one of the main input pathways to the cerebellum.

Q 5-40 An hour after an uncomplicated transection of the spinal cord at the level of C8 in an adult human:
(a) there is profuse sweating of the skin of the trunk.
(b) there is a spastic paralysis in the limbs.
(c) the bladder is atonic.
(d) the patient is unconscious.
(e) the blood pressure is elevated.

Q 5-41 The long-term consequences of hemisection of the spinal cord at the level of T8 include:
(a) an ipselateral extensor plantar response (Babinski positive).
(b) ipselateral spasticity below the level of the lesion.
(c) loss of temperature sense in the ipselateral leg.
(d) loss of position sense in the ipselateral leg.
(e) loss of crude touch sense in the ipselateral leg.

Q 5-42 Long-term consequences of uncomplicated transection of the spinal cord at the level of C8 include:
(a) loss of tendon jerks in the legs.
(b) loss of abdominal reflexes.
(c) loss of cremasteric reflex.
(d) flexor plantar responses.
(e) loss of the normal shivering reflex in the legs in response to a cold environment.

Q 5-43 Long-term consequences of uncomplicated transection of the spinal cord at the level of C8 in an adult man include:
(a) lack of reflex micturition.

(b) reflex fall in blood pressure as the bladder fills.
(c) loss of reflex erection of the penis.
(d) orthostatic hypotension.
(e) lack of sweating below the level of the lesion.

Q 5-44 The Purkinje cells of the cerebellar cortex have axons which:
(a) constitute the main efferent pathway from the cerebellar cortex.
(b) may terminate in the spinal cord.
(c) may terminate in the cerebellar nuclei.
(d) may terminate in excitatory synapses.
(e) produce influences mainly on ipselateral musculature.

Q 5-45 In each case, does a lesion in the structure mentioned on the left give rise to the condition mentioned on the right?

(a) Cerebral cortex	Tremor at rest.
(b) Cerebellum	Tremor appearing as an accompaniment of voluntary movement.
(c) Cauda equina	Dorsiflexion in Babinski's test.
(d) Medulla	Nasal regurgitation during swallowing.
(e) Pyramidal tract	Exaggerated knee-jerks.

Q 5-46 Damage to the basal ganglia usually causes in limb musculature:
(a) spasticity.
(b) interference with voluntary movement.
(c) involuntary movements.
(d) intention tremor.
(e) disorder of posture.

Q 5-47 With respect to brain structure and function:
(a) if a patient becomes anoxic, the cerebral cortex suffers irreversible damage before the brain stem.
(b) disturbance of consciousness can result from damage confined to the brain stem.
(c) the sleep centres lie in the basal ganglia.
(d) in the cerebral cortex of a normal subject, the two hemispheres are of approximately equal importance in speech.
(e) a lesion of Broca's area usually results in excessive talking.

Q 5-48 If the cerebral cortex is irrevocably damaged by a period of anoxia but the patient is still breathing spontaneously:

(a) the limbs will withdraw from a painful stimulus.
(b) the pupils will be unresponsive to light.
(c) the patient may speak.
(d) the limbs will have normal muscle tone.
(e) the eyes may be open and the eyeballs moving.

Q 5-49 The hypothalamus contains cells which are sensitive to:

(a) arterial blood pressure.
(b) TSH concentration.
(c) hydrogen ion concentration.
(d) partial pressure of oxygen.
(e) plasma volume.

Q 5-50 Concerning the nerve supply to the human skin:

(a) the promotion of heat loss from the skin involves increased activity in some sympathetic nerve fibres and decreased activity in others.
(b) the cold clammy skin of a patient in haemorrhagic shock is due to intense activity in all the sympathetic nerve fibres to the skin.
(c) in man, the sympathetic postganglionic fibres supplying sweat glands are cholinergic.
(d) the forearm skin receives a parasympathetic nerve supply.
(e) the synapse between sympathetic preganglionic and postganglionic fibres lies in the dermis.

Q 5-51 Atropine blocks the action of acetylcholine at parasympathetic postganglionic nerve endings. Its local application to the eye causes:

(a) dilation of the pupil.
(b) impaired ability to focus on nearby objects.
(c) difficulty in looking upwards.
(d) a tendency towards impairment of fluid drainage from the anterior chamber.
(e) everything to appear dimmer than normal to the subject.

Q 5-52 Interruption of the cervical sympathetic trunk results in ipselateral:

(a) ptosis.
(b) pupillary dilatation.
(c) exophthalmos.

(d) vasodilatation in the skin of the face.
(e) paralysis of lacrimation.

Q 5-53 Concerning the autonomic nervous system:
(a) parasympathetic nerve fibres usually have a synaptic relay close to or in the structure which they innervate.
(b) the sympathetic nerve supply to the hand arises from the thoracic cord.
(c) most of the cell bodies of sympathetic nerve fibres leaving the spinal cord lie in the dorsal root ganglia.
(d) the presynaptic sympathetic nerve fibres are thicker than the postsynaptic ones.
(e) at the synapse between pre-and postsynaptic sympathetic nerves, the transmitter is acetylcholine.

Section 6 Homeostasis

Q 6-01 Compared with intracellular fluid, extracellular fluid has:
(a) a greater osmolarity.
(b) a lower sodium ion concentration.
(c) a lower chloride ion concentration.
(d) a lower potassium ion concentration.
(e) a lower hydrogen ion concentration.

Q 6-02 With reference to extracellular fluid volume and composition in the human:
(a) if blood volume is depleted, adjustments involve an increase in aldosterone secretion.
(b) if the body is short of water, the collecting ducts of the nephrons become less permeable to water.
(c) plasma sodium concentration is regulated by varying the fraction of the sodium in the glomerular filtrate which is reabsorbed in the proximal tubules.
(d) increased osmolality leads to an increase in the secretion of anti-diuretic hormone (ADH).
(e) there is no absorption of sodium in the intestinal tract when the plasma sodium concentration is normal.

Q 6-03 With reference to water and solutes:
(a) diffusion means that molecules move in one direction only from a region of greater to a region of lesser concentration.
(b) the oncotic (colloid osmotic) pressure of the plasma represents about one-tenth of the total plasma osmolality.

(c) in plasma, Cl^- is in higher concentration than any other anion.
(d) the sum of the concentration of sodium and chloride ions on one side of a membrane permeable to these ions, must equal the sum of the two on the other side.
(e) carrier-mediated mechanisms are more susceptible to temperature changes than are diffusional processes.

Q 6-04 If about 1/2 litre of isotonic saline (sodium chloride solution) were infused intravenously in a healthy adult, consequences would include:
(a) increase in cardiac stroke volume.
(b) increase in flow of lymph from peripheral tissues.
(c) increase in renin secretion by the kidneys.
(d) increase in cerebral blood flow.
(e) equal distribution of the excess volume between intracellular and extracellular compartments.

Q 6-05 If an excessive amount of water is drunk:
(a) only a small fraction of it is absorbed from the gastro-intestinal tract.
(b) the osmolality of interstitial fluid decreases.
(c) receptors in the right atrium may be stimulated.
(d) receptors in the carotid body may be stimulated.
(e) an increase in glomerular filtration rate is the main means of disposing of the extra water.

Q 6-06 Concerning plasma albumin:
(a) less than 10% is degraded each week.
(b) its molecular weight is greater than that of globulin.
(c) it is responsible for most of the colloid osmotic pressure (oncotic pressure).
(d) in normal subjects, the albumin : globulin ratio is about 2 : 1 (weight for weight).
(e) a reduction in plasma albumin concentration is liable to be associated with oedema.

Q 6-07 Cerebrospinal fluid:
(a) is more acid than blood plasma.
(b) would be formed at a greater rate if the plasma became hypertonic.
(c) is formed by simple ultrafiltration of plasma.

(d) contains less protein than plasma.
(e) is absorbed into the venous sinuses.

Q 6-08 Concerning the cerebrospinal fluid:
(a) about half of it is formed at the choroid plexuses.
(b) it is in free ionic communication with the extracellular fluid compartment of the central nervous system.
(c) it contains a lower potassium concentration than blood plasma.
(d) ventilation is stimulated if its pH is decreased.
(e) the pH of the CSF is readily changed if plasma [H^+] alters.

Q 6-09 With reference to water in the body:
(a) the total accounts for approximately 60% of body weight in a lean adult man.
(b) approximately two-thirds of the total volume is intracellular.
(c) the total volume may be estimated using radioactively labelled plasma albumin.
(d) there is inevitably a continuous loss of water via the skin under average temperature conditions.
(e) ingestion of an unusually large quantity of water results in a transient decrease in plasma osmolarity.

Q 6-10 With reference to potassium in the body:
(a) the equivalent of most of the daily dietary intake of potassium is secreted into the tubular fluid in the kidneys.
(b) an increase in extracellular potassium concentration endangers cardiac function.
(c) a decrease in extracellular potassium concentration endangers cardiac function.
(d) an increase in extracellular potassium concentration (e.g. to 8 mmol/l) relaxes vascular smooth muscle.
(e) deficiency of aldosterone secretion results in an increase in extracellular potassium concentration.

Q 6-11 Normal renal tubular function includes:
(a) formation of ammonia.
(b) reabsorption of about 50% of the water of the glomerular filtrate.
(c) reabsorption of all the glucose in the glomerular filtrate.
(d) active transport of urea.
(e) formation of bicarbonate.

Q 6-12 In the kidney:
(a) half of the plasma flowing to the glomerular capillaries is filtered into the tubules.
(b) all of the sodium which passes into the glomerular filtrate is normally reabsorbed in the proximal tubule.
(c) acidaemia results in increased ammonia synthesis in tubular cells.
(d) the counter current mechanism results in the tubular fluid, when it leaves the loop of Henle, having a higher osmolarity than plasma.
(e) the excretion of an excessive water intake is assisted by decreased permeability to water of the collecting ducts.

Q 6-13 At the renal glomeruli, in normal physiological conditions:
(a) fluid in the Bowman's capsule has the same electrolyte concentrations as plasma.
(b) blood in the efferent arterioles is more viscous than blood in the afferent arterioles.
(c) the hydrostatic pressure in the capillaries varies as the arterial blood pressure varies.
(d) the hydrostatic pressure in the capillaries is higher than in capillaries elsewhere, in a supine subject.
(e) the glucose concentration in the plasma leaving the glomerulus is virtually the same as that in the plasma entering the glomerulus.

Q 6-14 The following measurements were made on a subject:

plasma glucose concentration	18.0 mmol/l
urinary excretion of glucose	0.18 mmol/min
glomerular filtration rate	125 ml/min
urine flow	2 ml/min.

Does this information allow the following conclusions to be drawn?
(a) The urine flow indicates that a diuresis is occurring.
(b) The urine concentration of glucose is 75 mmol/l.
(c) The renal clearance of glucose is 10 ml/min.
(d) The T_m (transport maximum) for glucose is between 2.0 and 2.1 mmol/min.
(e) The data indicate that glucose is actively secreted into the urine.

Q 6-15 In relation to renal function:
(a) renal plasma clearance of a substance is the volume of plasma which contains the amount of the substance appearing in the urine per minute.

(b) a substance which is filtered at the glomerulus, but is neither secreted nor reabsorbed in the tubules, will have a clearance value equal to the volume of plasma that flows through the renal circulation in one minute.
(c) the clearance of inulin is the volume of glomerular filtrate produced per minute.
(d) it is possible for the osmolarity of the urine to exceed the osmolarity of the interstitial fluid in the tip of the renal medullary papillae.
(e) correction of an increase in extracellular fluid volume includes an increase in secretion of aldosterone.

Q 6-16 With respect to plasma clearance and renal function:

(a) the clearance of a substance can exceed the subject's glomerular filtration rate (GFR) only if the substance is secreted into the tubular fluid.
(b) if a substance is secreted into the urine, its clearance must exceed that of the subject's glomerular filtration rate.
(c) the renal clearance for glucose in a normal subject is usually less than 1 ml/min.
(d) if the plasma glucose concentration rises to three times normal (as a result, for instance of glucose injected intravenously), the glucose clearance will rise.
(e) if the plasma bicarbonate concentration doubles, the renal clearance of bicarbonate will rise.

Q 6-17 If the concentration of a substance A in the plasma is Pa (mg/ml) and in the urine is Ua (mg/ml), and the volume of urine produced per minute is V (ml), are the following statements correct?

(a) The rate of excretion of substance A is Ua × V (mg/min).
(b) The quantity Ua × V/Pa is the minimum volume of plasma from which the kidneys could have obtained the amount of A excreted per minute.
(c) If A is filtered at the glomerulus and neither secreted nor absorbed in the renal tubules, then Ua × V/Pa is the volume of plasma that flows through the renal circulation in one minute.
(d) If A is inulin then Ua × V/Pa is the volume of glomerular filtrate produced per minute.
(e) If the ratio Ua/Pa exceeds the ratio Ub/Pb for a substance B which is only filtered, then A must be secreted by the renal tubules.

Q 6-18 The following data concerning the renal handling of sodium and potassium are from a normal subject.

	Na^+ (mmol)	*K^+ (mmol)*
Amount per 24 hours:		
filtered in glomeruli	26 000	900
reabsorbed	25 850	900
secreted	0	100

In this subject:

(a) the amount of sodium appearing in the urine is less than 5% of the filtered load.

(b) the amount of potassium appearing in the urine is less than 5% of the filtered load.

(c) the urinary loss of sodium exceeds that of potassium by a factor in excess of 10.

Concerning plasma electrolyte concentrations in general:

(d) the concentration of sodium in plasma exceeds that of potassium by more than tenfold.

(e) the principal plasma anion is bicarbonate.

Q 6-19 With regard to the mechanisms leading to the concentration of urine:

(a) the urine entering the descending limb of the loop of Henle is approximately isotonic with arterial plasma.

(b) the descending limb of the loop of Henle is freely permeable to electrolytes.

(c) tubular fluid in the thin segment of the ascending limb is approximately isotonic with arterial plasma.

(d) the thick portion of the ascending limb pumps electrolytes into the tubular fluid from the extracellular fluid.

(e) the concentration of urea in the extracellular fluid of the renal medulla is greater than in that of the renal cortex.

Q 6-20 With reference to the renal tubules:

(a) at the tip of the loop of Henle in the renal medulla, the osmolarity of the tubular contents is several times that of the glomerular filtrate.

(b) the walls of the ascending limb of the loop of Henle are freely permeable to water.

(c) the concentration of creatinine in the tubular fluid increases with distance along the tubule.

(d) most of the water filtered at the glomerulus is reabsorbed in the distal convoluted tubules and collecting ducts.

(e) the permeability of the collecting tubule to water is under the control of aldosterone.

Q 6-21 In a normal subject excreting acid urine:

(a) hydrogen ion movement from the tubular cytoplasm to the luminal fluid in the collecting ducts is down the electrochemical gradient for hydrogen ions.

(b) the minimum urinary pH which can be attained is about 6.

(c) the rate of hydrogen ion secretion in the proximal tubule is greater than in the distal tubule.

(d) a higher hydrogen ion concentration difference between tubular fluid and tubular cell cytoplasm is achieved in the proximal than in the distal tubule.

(e) hydrogen ion secretion occurs largely in exchange for the reabsorption of chloride ions.

Q 6-22 Concerning acid-base homeostasis and the kidney:

(a) the renal clearance of bicarbonate may be 20 times higher in a vegetarian than in a subject on a mixed diet.

(b) the ingestion of sodium lactate will result in an increase in renal excretion of bicarbonate.

(c) urinary ammonia is largely derived from urea.

(d) the ammonium ions in the tubular fluid originate from molecular ammonia transferred across the tubular epithelial cell membrane.

(e) acidosis is likely to be accompanied by an increased concentration of ammonium ions in the urine.

Q 6-23 In a patient suffering from chronic vomiting of gastric juice:

(a) the pH of the extracellular fluid is likely to be raised.

(b) the plasma bicarbonate concentration is likely to be raised.

(c) the plasma potassium concentration is likely to be raised.

(d) the plasma chloride concentration is likely to be raised.

(e) the urine may be acid.

Q 6-24 With reference to extracellular fluid (ECF) hydrogen ion concentration *in vivo*:

(a) the normal value is near to $10^{-4.4}$ mmol/l.

(b) to maintain normality on an average diet, the kidney must excrete about 50 to 60 mmol of H^+ daily.
(c) when ECF acidity increases, ventilation is stimulated.
(d) when ECF acidity increases, arterial PCO_2 increases.
(e) when ECF acidity increases, less bicarbonate is generated in renal tubular cells.

Q 6-25 Concerning acid-base balance:
(a) plasma proteins are quantitatively more important than plasma phosphate in buffering.
(b) the renal hydrogen ion secretion mechanism is transport maximum (Tm) limited.
(c) when the arterial PCO_2 rises, the amount of bicarbonate reabsorbed by the tubules is increased.
(d) bicarbonate reabsorption in the renal tubules is largely dependent on the tubular hydrogen ion pump mechanism.
(e) as a result of the bicarbonate reabsorption mechanism, the body rids itself of excess acidity in the form of water.

Q 6-26 The Henderson-Hasselbalch equation applied to the CO_2/bicarbonate equilibrium at 37°C is $pH = 6.1 + \log([HCO_3^-])/(0.03\ PCO_2)$ where PCO_2 is measured in mmHg and $[HCO_3^-]$ is in mmol/l. ($\log 2 = 0.30$):
(a) if $[HCO_3^-] = 24$ mmol/l and $PCO_2 = 40$ mmHg, then the pH is 7.3.
(b) if the pH falls by 0.3 of a unit with a constant PCO_2, the $[HCO_3^-]$ must fall by one half.
(c) in uncompensated respiratory acidosis, the plasma $[HCO_3^-]$ rises.
(d) in uncompensated respiratory acidosis, it is the bicarbonate buffer system which limits the change in hydrogen ion concentration.
(e) in metabolic acidosis the plasma bicarbonate concentration tends to be lowered.

Q 6-27 Concerning acid-base balance and body fluids:
(a) the amount of undissociated carbonic acid in an aqueous solution is inversely proportional to the amount of dissolved carbon dioxide.
(b) the amount of undissociated carbonic acid is approximately equal to the amount of dissolved carbon dioxide at a PCO_2 of 40 mmHg.

(c) a rise in plasma carbon dioxide concentration associated with a proportional rise in bicarbonate ion concentration is accompanied by a fall in pH.
(d) the renal compensation for a respiratory acidosis includes an increase in the rate of addition of bicarbonate to the blood by the renal tubules.
(e) as a chemical buffer in a closed system, the bicarbonate buffer system is more efficient at a pH of 7.1 than at a pH of 7.4.

Q 6-28 A patient with fully compensated respiratory acidosis has:
(a) a normal arterial PCO_2.
(b) a normal plasma pH.
(c) a normal plasma bicarbonate concentration.
(d) a normal bicarbonate:PCO_2 ratio in plasma.
(e) a negative base excess.

Q 6-29 With respect to acute respiratory disturbances of acid base balance:
(a) retention of carbon dioxide, due to respiratory depression, increases the acidity in all body fluid compartments.
(b) in either type of respiratory disturbance, CSF pH will change in the same direction as blood pH.
(c) in persistent hyperventilation, there will be a low rate of bicarbonate reabsorption in the kidneys.
(d) in brief vigorous hyperventilation, cerebral blood vessels constrict.
(e) when arterial PCO_2 rises, cerebral vasodilatation can help to protect against excessive tissue acidity in the brain.

Q 6-30 With reference to the control of extracellular fluid volume and osmolarity:
(a) aldosterone acts in the kidney by increasing the permeability of the distal tubules to water.
(b) renin is secreted by cells situated close to the afferent glomerular arterioles.
(c) antidiuretic hormone is released in response to the action of hypothalamic releasing factors.
(d) an increase in right atrial pressure leads to a decrease in aldosterone activity.
(e) an excessive secretion of aldosterone results in increased loss of potassium in the urine.

Q 6-31 Are the values given appropriate for a healthy 70-kg man?
(a) the body contains about 40 litres of water.

(b) the minimal daily water intake to maintain fluid balance in temperate conditions is approximately 450 ml.
(c) a suitable daily calorie intake for moderate physical activity would be 3500 kcal.
(d) the glomerular filtration rate (both kidneys) is approximately 1200 ml/min.
(e) Plasma osmolarity is about 300 mOsmol/litre.

Q 6-32 With reference to endocrine function:
(a) growth hormone is involved in the control of growth of the long bones during adolescence.
(b) cortisol concentration in the plasma exerts a negative feedback effect on hypothalamic production of corticotrophic releasing factor.
(c) steroid hormones are normally bound to plasma proteins.
(d) adrenal androgens are normally secreted only in the male.
(e) release of antidiuretic hormone from the posterior pituitary is increased when osmolality of cerebral extracellular fluid increases.

Q 6-33 With reference to variations in plasma constituents:
(a) plasma sodium concentration rises when aldosterone secretion is deficient.
(b) the concentration of calcium ions in the plasma is decreased by vigorous hyperventilation.
(c) arterial carbon dioxide tension ($PaCO_2$) is normal in compensated respiratory alkalosis.
(d) arterial oxygen tension (PaO_2) is below normal in anaemia.
(e) plasma pH is low during exhausting muscular work.

Q 6-34 In the following situations, each involving feedback, decide whether the described effect is part of a NEGATIVE feedback system.
(a) The secretion of secretin into the blood in response to acidity of the duodenal contents.
(b) The opening of sodium channels in the nerve membrane in response to depolarization of the membrane.
(c) The secretion of bile salts by the liver into the hepatic bile in response to bile salts in the plasma.
(d) The release of TSH (thyroid stimulating hormone) in response to a low plasma concentration of thyroxine.

(e) The secretion of LH (luteinizing hormone) just before ovulation in response to a rise in circulating oestrogen levels.

Q 6-35 The hypothalamus:

(a) contains the pneumotaxic centre.
(b) contains receptors for blood temperature.
(c) synthesises hormones.
(d) is linked neurally with the posterior pituitary.
(e) influences the anterior pituitary secretions by way of a 'portal' system of connecting blood vessels.

Q 6-36 With reference to the mechanisms of regulation of concentrations in the plasma of the several substances:

(a) a fall in blood glucose is corrected mainly by replacement from muscle glycogen.
(b) a rise in blood urea is corrected by a rise in the filtered load of urea in the glomeruli.
(c) a rise in blood glucose is partly corrected by lowering the renal tubular maximum (Tm) for reabsorption.
(d) a rise in plasma Ca^{2+} leads to greater deposition of calcium salts in bone.
(e) plasma albumin is replenished by the liver.

Q 6-37 With reference to sensors for homeostatic regulatory mechanisms:

(a) hypothalamic receptors sense alterations in extracellular osmolality.
(b) the carotid sinus receptors sense alterations in arterial PO_2.
(c) atrial wall receptors sense changes in central venous pressure.
(d) receptors in the liver account for the response which corrects alterations in blood glucose.
(e) the need for adjustment of sodium reabsorption is sensed in the kidneys.

Q 6-38 Does the condition on the left lead to an increase in secretion of the hormone on the right?

(a) Persistent exposure to cold	Thyroxine
(b) Dietary deficiency of iodine	TSH
(c) Increase in blood glucose concentration	Glucagon
(d) Excessive water intake	Renin

(e) Dilatation of the cervix uteri in labour — Oxytocin

Q 6-39 With reference to features of endocrine function in general:

(a) all hormones reach their targets by means of dispersion throughout the whole circulation.
(b) the binding of a hormone to receptor sites on target cells increases in direct proportion to the plasma concentration of the hormone.
(c) hormones which are fat-soluble and of relatively low molecular weight may enter cells by diffusion before acting in a specific manner.
(d) many hormones act by altering the rate at which some substance is transported across the cell membrane of the target cells.
(e) radioimmunoassay is a method of estimating the amount of antibody produced when a particular hormone is injected.

Q 6-40 With reference to hormones:

(a) the concentration of a hormone bound to plasma protein is normally greater than the concentration of its free form.
(b) plasma concentration of a hormone depends essentially on its secretion rate since inactivation and excretion rate vary relatively little.
(c) the molecular weight of all hormones in the free form is such that they are too large to be lost in the glomerular filtrate.
(d) thyroid stimulating hormone (TSH) is secreted by an autotransplanted anterior pituitary gland.
(e) prolactin is secreted by an autotransplanted anterior pituitary gland.

Q 6-41 With reference to thyroid function:

(a) iodine deficiency results in enlargement of the thyroid gland.
(b) thyroxine is carried in the blood as thyroglobulin.
(c) tri-iodothyronine is more biologically active than tetraiodothyronine (thyroxine).
(d) if the binding protein concentration rises in the plasma, the concentration of thyroid hormone in the free state will decrease.
(e) an increase in tissue oxygen usage is caused by thyroid hormone.

Q 6-42 With reference to the secretions of the anterior pituitary:

(a) the anterior pituitary contains much more growth hormone than any other hormone.
(b) deficiencies produced by removal of the anterior pituitary could be corrected by reimplanting it at some other site in the body.
(c) a fall in plasma glucose concentration leads to an increase in growth hormone secretion.
(d) follicle stimulating hormone (FSH) is secreted by the anterior pituitary only in females.
(e) the period of peak prolactin secretion is around the time of parturition.

Q 6-43 Injection of parathyroid hormone leads to:

(a) increase in urinary phosphate.
(b) increased production of dihydroxycholecalciferol in the kidney.
(c) an increase in the number of osteoblasts.
(d) a rise in plasma calcium.
(e) increased reabsorption of calcium in the kidney.

Q 6-44 A low serum calcium may be associated with:

(a) hyperexcitability of peripheral nerves.
(b) reduced calcitonin release.
(c) an increased serum phophate concentration.
(d) vitamin D deficiency.
(e) reduced parathormone release.

Q 6-45 With reference to calcium:

(a) calcitonin can start to correct a rise in plasma calcium level within a few minutes.
(b) calcium is released from smooth muscle cells when they contract.
(c) calcium is necessary for blood clotting.
(d) parathormone decreases calcium absorption from the gut.
(e) calcium ion concentration in interstitial fluid is about 1.2 mmol/l.

Q 6-46 With respect to calcium in man:

(a) there is a higher concentration of free calcium ions inside cells than outside.
(b) bound calcium in the plasma dissociates more when pH decreases.
(c) low plasma calcium concentration leads to the release of more 1.25-DHCC.

(d) phytic acid in the diet diminishes absorption of calcium from the gut.
(e) a minimum daily dietary requirement in an adult is about 100 mg.

Q 6-47 In bone in the mature person:
(a) three-quarters of the substance is mineral, one-quarter organic.
(b) bed rest can deplete the mineral content.
(c) osteoblasts are associated with areas where resorption is taking place.
(d) alkaline phosphatase is associated with the laying down of mineral.
(e) calcium is transferred to the extracellular fluid in response to an increase in calcitonin.

Q 6-48 During the normal ovarian cycle:
(a) ova multiply during the first half of the cycle.
(b) a single mature Graafian follicle discharges its oocyte about the middle of the cycle.
(c) oestradiol secretion increases during the first half of the cycle.
(d) progesterone secretion inhibits FSH secretion.
(e) a corpus luteum develops only if the ovum is fertilized.

Q 6-49 During the normal menstrual cycle:
(a) the endometrium proliferates in the first half of the cycle.
(b) the endometrium secretes gonadotrophins in the second half of the cycle.
(c) a decrease in secretion of both oestrogen and progesterone at the end of the cycle brings about the breakdown of the endometrium.
(d) the mucus secretion of the cervix becomes thicker at mid-cycle.
(e) there is a rise in body temperature at mid-cycle.

Q 6-50 During the first three months of pregnancy:
(a) chorionic gonadotrophin acts to maintain the corpus luteum.
(b) prolactin is secreted in increasing amounts by the maternal anterior pituitary gland.
(c) the smooth muscle of the uterus is relatively inexcitable because of the action of progesterone.
(d) pregnancy can be diagnosed by means of tests based on the presence in the maternal urine of pituitary gonadotrophins.

(e) there is a progressive reduction in the total vascular peripheral resistance.

Q 6-51 The secretion of testosterone from the testis:
(a) starts during fetal life.
(b) causes the development of the external genitalia.
(c) stimulates androgen secretion from the adrenal cortex.
(d) is controlled by the male equivalent of luteinizing hormone.
(e) has no influence on spermatogenesis.

Q 6-52 When a healthy individual is exposed to cold:
(a) metabolic heat production is increased by catecholamine secretion.
(b) the total peripheral vascular resistance decreases.
(c) surface temperature decreases more than core temperature.
(d) adjustments are controlled from centres in the medulla.
(e) shivering further decreases the body temperature.

Q 6-53 A survivor of a shipwreck with a core temperature of 30°C:
(a) has a higher metabolic rate than a normal subject at rest.
(b) shivers violently.
(c) is unconscious.
(d) has a decreased blood pressure.
(e) has a raised peripheral resistance.

Q 6-54 If a normal individual is exposed to a hot environment:
(a) the additional heat loss through sweating depends on evaporation of the sweat.
(b) a reflex increase in ventilation is a major way of increasing heat loss.
(c) if sweating is excessive, salt intake should be cut down.
(d) in the process of maintaining water balance, urine output can decrease to about 200 ml per day.
(e) resting cardiac output will be higher than in his usual environment.

Q 6-55 With reference to the maintenance of body temperature in the human:
(a) during prolonged exposure to hot environments, the metabolic rate increases.

(b) all mechanisms controlling heat loss from the skin surface are served by noradrenergic sympathetic nerves.
(c) changes in temperature of the skin are more potent than changes in temperature of the hypothalamus in evoking homeostatic responses.
(d) when the environmental temperature is high, a high relative humidity facilitates the loss of heat by the subject.
(e) in a fever of rapid onset when the core temperature is rising, the subject is likely to feel cold.

Q 6-56 Concerning body temperature and its control in the human:

(a) evaporation of water from the body surface occurs only when the sweat glands secrete.
(b) if the core temperature rises, the subject will sweat even if the fluid loss results in circulatory collapse.
(c) in a trained athlete, the sodium concentration is the same in plasma and sweat.
(d) the loss by a healthy adult of 4 litres of sweat during a working shift in a hot environment endangers life.
(e) a core temperature of 43°C signifies a danger to life.

Answers

Section 1 Digestion, Absorption and Metabolism

A 1-01

(a) **No** Carbohydrate digestion starts in the mouth: salivary amylase (ptyalin) acts on starch: this amylase activity is inhibited by gastric acidity.

(b) **Yes** Final breakdown occurs at the brush border of the epithelial cells, where the appropriate enzymes are located.

(c) **Yes** Monosaccharides are the final product of digestion, and these are absorbed.

(d) **No** An active carrier mechanism is involved; inhibition of sodium pump activity interferes with absorption; different sugars have different absorption rates. Glucose is absorbed down a diffusion gradient when there is one, but this is not necessary for absorption.

(e) **No** Fats have more than twice the energy value (39 kJ as compared to 17 kJ, or about 9 : 4 kcal, per g).

A 1-02

(a) **No** Three hours is about right, although there is variation with the nature of the food.

(b) **No** Vagal activity enhances, sympathetic depresses.

(c) **No** There is 'receptive relaxation' as food enters, so that pressure is unchanged.

(d) **Yes** This produces waves of contraction, every 20 seconds.

(e) **No** Histamine stimulates acid secretion.

A 1-03

(a) **Yes** This is the 'alkaline tide': extrusion of H^+ into the lumen is associated with a rise in HCO_3^- concentration in cells: this moves out into plasma.

(b) **No** Digestion is assisted by gastric acidity, but no vitally necessary process is impossible without it.

(c) **Yes** Pepsin is split off only when pH falls (acidity increases) below about 5.

(d) **No** Intrinsic factor is a necessary adjunct to the Vitamin B_{12} which comes from the diet. Without intrinsic factor, B_{12} is not absorbed and pernicious anaemia ensues.

(e) **Yes** In the cephalic phase, vagal stimulation accounts for pepsinogen and acid secretion, directly by stimulation of secretory cells and indirectly by causing release of gastrin.

A 1-04

(a) **No** The correct value is between 1 and 2.

(b) **Yes** The gastric receptors are H-2 receptors. The standard histamine blockers such as anthisan block pulmonary histamine receptors (H-1) but not the receptors in the stomach.

(c) **Yes** This is the standard means of eliciting a maximal secretion of acid and thus assessing the 'parietal cell mass'.

(d) **Yes**

(e) **No** Gastrin is secreted by G cells located in the antrum.

A 1-05

(a) **No** They are polypeptides.

(b) **No** Secretin is secreted by the duodenal mucosa.

(c) **Yes**

(d) **Yes**

(e) **Yes** About 1½ litres in 24 hours.

A 1-06

(a) **No** Secretin is the important hormone. (see 1–05(c))

(b) **No** Chloride ions are reabsorbed from the bile into the interstitial fluid in exchange for bicarbonate ions.

(c) **Yes**

(d) **No** This would be true of bile salts but not of bile pigments, only a tiny proportion of which are reabsorbed.

(e) **No** Dietary intake is usually around 1.5 litres per day. Secretions into the gastrointestinal tract total at least 10 times this amount.

A 1-07

(a) **Yes** Small volume, enzyme-rich fluid is secreted in the 'cephalic phase'.

(b) **Yes** Copious water and electrolyte-rich secretion is the response to secretin.

(c) **Yes** The effect is muscarinic cholinergic.
(d) **No** Secretin acts directly on the pancreas.
(e) **Yes** CCK-PZ elicits enzyme containing secretion, (though it does also enhance the HCO_3 secretion in response to secretin).

A 1-08
(a) **Yes**
(b) **Yes** This causes inadequate bile salt secretion and hence deficient fat absorption.
(c) **Yes** as (b).
(d) **Yes** Inadequate lipase secretion results in deficient fat digestion.
(e) **No** The stomach does not contribute significantly to fat digestion and absorption.

A 1-09
(a) **Yes**
(b) **No** This occurs in the liver.
(c) **No** It contains no iron.
(d) **No** Because it is bound to plasma protein.
(e) **No** This product of haemoglobin breakdown cannot be recycled by the body.

A 1-10
(a) **Yes** But bile remains isosmotic with plasma partly because of micelle formation.
(b) **No** The reverse is true since bile salts render cholesterol soluble.
(c) **Yes** This is also the route of elimination when plasma cholesterol is elevated.
(d) **Yes**
(e) **No** 10% indicates malabsorption; the figure for a healthy human is about 1%.

A 1-11
(a) **No** This would be true for bile pigments.
(b) **No** They are secreted into the bile canaliculi in the liver.
(c) **Yes** Because sodium and water are absorbed.
(d) **Yes**
(e) **No** This is the function of lipases; bile salts assist the breakdown by taking part in emulsification and micelle formation.

A 1-12
(a) **Yes**
(b) **No** The postganglionic fibre originates at a synapse within the wall in the myenteric plexus.

(c) Yes
(d) Yes and relays in the coeliac or superior mesenteric ganglia.
(e) No They are noradrenergic.

A 1-13
(a) Yes Most of this is from gut secretions.
(b) No About half the ingested protein is absorbed as amino acids.
(c) Yes There are different carrier systems for the different groups of substances.
(d) No Most absorption of vitamin B_{12} occurs in the ileum.
(e) Yes Some is absorbed into the lymphatics and passes thence straight to the systemic circulation.

A 1-14
(a) No Glucose is absorbed mostly in the proximal small gut.
(b) No The jejunum is the principal site for the absorption of the products of fat digestion.
(c) Yes
(d) No Most vitamin K is absorbed in the large bowel after bacteria synthesize it in the lumen of that region.
(e) No Most iron is absorbed in the jejunum.

A 1-15
(a) Yes Secretion is essentially mucous.
(b) No Sympathetic stimulation decreases colonic motility.
(c) Yes This is one mechanism of emotional diarrhoea.
(d) Yes Water is absorbed consequent on active sodium reabsorption.
(e) Yes Dietary intake is normally insufficient.

A 1-16
(a) No There is no evidence that the adverse effects of constipation are due to factors other than abdominal discomfort and distension.
(b) Yes Distension of the rectum stimulates receptors; via sacral segments of the spinal cord, parasympathetic (nervi erigentes) efferents cause contraction of rectal, and relaxation of internal sphincteric, smooth muscle. This is superimposed on a weaker local reflex.

(c) **Yes** This is the so-called gastrocolic reflex. Distension of the stomach is frequently accompanied by an urge to defaecate.

(d) **Yes** —see (b).

(e) **Yes**

A 1-17

(a) **No** Vestibular receptors are primarily involved.

(b) **No** It promotes vomiting.

(c) **No** The vomiting centre is in the medulla.

(d) **No** The duodenum contracts whilst the stomach relaxes in this situation.

(e) **Yes** For example the parasympathetic component of the autonomic nervous system is responsible for stomach contraction and the somatic nervous system for contraction of the abdominal musculature.

A 1-18

(a) **Yes** Blood glucose is maintained at this level even in prolonged fasting, by liver gluconeogenesis, from amino acids derived in turn from muscle protein breakdown (after liver glycogen is depleted within first day).

(b) **Yes** Insulin secretion is regulated by the level of blood glucose, so it decreases and this prevents uptake of glucose in muscle—but *not* in brain.

(c) **Yes** Lipolysis in fat stores releases FFA which are taken up by muscle.

(d) **Yes** Liver gluconeogenesis is promoted by decrease in insulin and increase in glucagon.

(e) **No** Brain requires glucose and takes what is available (see (b)); it is also able to utilize keto-acids, after a few days, along with glucose.

A 1-19

(a) **No** It is normally back to normal within one and a half hours.

(b) **No** It is 'completely filtered', but only one-fifth of the plasma is filtered off.

(c) **Yes** This is the first source.

(d) **Yes** Brain normally uses only glucose; in starvation it can use keto-acids only if there is also some glucose.

(e) **No** It is in free solution.

A 1-20

(a) **Yes** Prothrombin is formed only in the liver, and its formation requires vitamin K.

(b) **No** RBCs disintegrate after about 4 months, and phagocytes in the reticulo-endothelial system (including the liver) take them up.

(c) **No** Other tissues (muscle) also convert glucose to glycogen.

(d) **Yes** Only liver possesses the appropriate enzyme for breakdown of glycogen to glucose. Muscle utilizes glycogen, breaking it down to lactate and pyruvate but does not release glucose into the blood.

(e) **Yes** Liver produces and secretes bile salts from bile acids, conjugated with glycine or taurine.

A 1-21

(a) **Yes** Hepatic artery flow is estimated to be about one-half that of portal vein flow: but the latter varies with intestinal activity.

(b) **Yes** The sum of the two inflows = outflow (hepatic vein) = about 1200–1400 ml/min, or about one-quarter cardiac output.

(c) **Yes** Flow decreases during exercise, probably by means of increased sympathetic outflow causing hepatic vasoconstriction; a fall in portal flow is secondary to a decrease in flow to the gut.

(d) **Yes** Hepatic arterioles and portal venules both open into the sinusoids, so that the liver cells are supplied with oxygen and also with the nutrients absorbed from the gut.

(e) **No** All types of foodstuff are represented in the portal vein, and it carries all of the protein and carbohydrate; but lymphatics carry some of the fat.

A 1-22

(a) **Yes** Hence cortisol tends to raise blood glucose.

(b) **No** Glucose enters without the help of insulin —otherwise, how would brain obtain its crucial supply, when blood glucose falls, and therefore insulin level falls also?

(c) **Yes** Insulin assists glucose to enter muscle cells, and promotes storage as glycogen.

(d) **Yes** A rise in glucagon, and fall in insulin, promotes restoration of blood glucose from gluconeogenesis in the liver.

(e) **Yes** This occurs in 'fight or flight'.

A 1-23

(a) **Yes** But only in very severe work: at lower work

levels, muscles use FFA and glucose from blood, depending on fibre type. Anaerobic glycolysis at very heavy loads utilizes local glycogen stores.

(b) **No** Blood glucose is maintained or increased implying greater production rate during the greater utilization, at first; later it may decrease.

(c) **Yes** Both direct sympathetic stimulation, and circulating catecholamines stimulate gluconeogenesis.

(d) **No** P_aCO_2 is kept normal initially, by appropriate increase in alveolar ventilation. It decreases when the anaerobic threshold is reached.

(e) **Yes** The anaerobic component of metabolism releases lactic acid, causing decreased plasma pH and increased plasma lactate.

Section 2 Breathing and Gas Exchange

A 2-01

(a) **No** The FRC is at least twice this volume.

(b) **No** When resistance to flow increases, the tidal excursion moves to a higher lung volume, so FRC is increased.

(c) **Yes** Helium is not taken up appreciably into the blood; a known concentration is rebreathed from a spirometer until there is a constant concentration in the mixed gas; thence total volume of lungs + spirometer can be estimated from the final concentration of He.

(d) **Yes** Because tidal volume is small relative to the volume remaining in the lungs continuously, there is very little fluctuation in CO_2 and O_2 concentrations in alveoli, and therefore in blood.

(e) **No** Airways begin to close only as residual volume is approached in healthy young people.

A 2-02

(a) **Yes** A typical total lung capacity might be 6 litres, and tidal volume 400 to 500 ml.

(b) **No** The mean volume around which tidal excursion takes place is at about mid lung volume.

(c) **Yes** In healthy young subjects, during forced expiration, airways in lower regions start to close just before RV is reached. Gravity causes intrapleural pressure to be less around the upper than around the lower parts of the lungs; the upper airways are therefore held open longer, as expiration proceeds.

(d) **Yes** At residual volume, airways in upper parts are held more open than those in lower parts; so the first air will enter the apices.

(e) **Yes** A complex mechanical feedback results in an appropriate compensation: when air is shifted in and out at a higher lung volume, there is more help during expiration from the elastic recoil.

A 2-03

(a) **Yes** The FEV_1 is 80% of the VC.

(b) **No** These could be perfectly normal values for a small woman: they are all in the correct proportions.

(c) **Yes** The ERV is found by subtracting RV from FRC.

(d) **No** It is not possible to measure RV by spirometer: a dilution method is required—a spirometer may be one component of the apparatus used.

(e) **Yes** FRC is the volume concerned. At the end of expiration, the lungs are full of alveolar gas, which has about 14% oxygen

$$0.14 \times 1.8 = 0.252 \text{ l}.$$

A 2-04

(a) **No** Surface tension is most strongly opposed by surfactant at low lung volume.

(b) **No** The part b-B shows a decreasing change in volume for a given change in pressure—the lung is becoming *less compliant* as it becomes full.

(c) **No** There would be hysteresis: that is, the path in the other direction—from full to residual volume—would be different (It would lie above and to the left of a-b).

(d) **Yes** Filling the lung with liquid abolishes the surface tension factor, expansion is easier, compliance is greater, a given change in pressure causes a larger change in volume.

(e) **No** A greater surfactant activity would make the lung more compliant. Curve A-D suggests a lack of normal surfactant activity.

A 2-05

(a) **Yes** Resistance to airflow during the breathing cycle affects the area of the pressure volume loop, and the slope of the line joining its ends.

(b) **Yes** This is the definition.

(c) **No** The system is most compliant in the normal tidal volume range.

(d) **Yes** Compliance is increased by surfactant, decreased by surface tension.

(e) **No** Alveoli at the apex, being already more stretched at end-expiratory volume due to gravity, are less easily filled than lower alveoli.

A 2-06

(a) **Yes** During inspiration, the elastic recoil of the lungs and surface tension necessitate muscular work; they assist expiration and at rest can account for all the work needed to overcome airflow resistance and chest compression.

(b) **Yes** During forced expiration the intrathoracic pressure becomes positive, and tends to close small airways, increasing the resistance to airflow.

(c) **Yes** The position of rest is above the tidal range.

(d) **Yes** There is more 'pull away' of the lungs from the chest wall at the top, so the pressure here is more negative.

(e) **No** They use only a tiny fraction: about 1 ml/min per litre/min which is of the order of 5 to 10 ml/min, or less than one-twenty-fifth of the whole body oxygen consumption.

A 2-07

(a) **No** Curve B–A, to the right, shows expiration.

(b) **Yes** This is a normal tidal breath, so inspiration starts from the functional residual capacity.

(c) **No** The volume change shown is 400 ml. $16 \times 0.4 = 6.4$ litres which is a possible vital capacity, but *not* a possible tidal volume in exercise. Oxygen consumption can increase by 16 times, and minute volume by a comparable factor; tidal volume does not usually become more than half the vital capacity.

(d) **Yes** Compliance is change in volume/change in pressure.

(e) **Yes** Resistance to airflow is one of the factors which makes it necessary to exert a greater pressure change to cause a given volume change, during inflation and deflation, than if the lungs were simply elastic structures. The greater the resistance, the further would the curves be from the straight line of compliance.

A 2-08

(a) **Yes** Surfactant diminishes the effect of surface tension in the alveoli and therefore limits the work which is required to expand them.

(b) **Yes** Work in expanding the lungs is done in part against the elastic recoil—the force tending always to deflate the lungs.

(c) **Yes** The compliance of the lungs—the ease with which they can be expanded—becomes less at volumes above the normal tidal range.

(d) **Yes** Bronchial dilatation leads to a decrease in resistance to airflow, against which work has to be done during inspiration (and expiration).

(e) **Yes** Despite the modifying effect of surfactant, there is a surface tension force which must be overcome in any increase in lung volume.

A 2-09

(a) **Yes** Ventilation = 8 litres/min. Stroke (= tidal) volume = 0.8 litres, therefore frequency = 10/min. Dead space = 150 ml, therefore dead space ventilation = 10×150 ml/min = 1.5 litres/min. Alveolar ventilation = total ventilation − dead space ventilation = $8 - 1.5 = 6.5$ litres/min.

(b) **Yes** Alveolar CO_2 = 3%: alveolar PCO_2 = 3/100 barometric pressure. $PaCO_2$ is less than 3 kPa (23 mmHg). By definition, this is hyperventilation.

(c) **No** Low PCO_2 causes cerebral vasoconstriction.

(d) **Yes** The inspired PO_2 will be about 50 kPa (380 mmHg). Alveolar PO_2 is roughly this minus alveolar PCO_2, which has been calculated as less than 3 kPa (23 mmHg). So the range is right for alveolar PO_2, and if pulmonary gas exchange is normal, arterial PO_2 will be not more than 1.5 kPa (10 mmHg) lower.

(e) **No** To answer this you need only know that RQ is expressed as CO_2 output ÷ O_2 uptake. 195/240 is clearly less than 1.

A 2-10

(a) **No** Hyperventilation only very slightly increases the oxygen consumption. So a smaller fraction of the oxygen is removed from the much larger volume of inspired gas. So the O_2% in expired gas increases.

(b) **Yes** By about 20 litres per min per litre O_2 per min.

(c) **Yes** In this method, the subject rebreathes oxygen, and the CO_2 is absorbed. Therefore the rate of decrease of volume is a direct measure of the O_2 consumed.

(d) **Yes** But varies, of course, according to body size.

(e) **Yes** This conversion factor is used to calculate the metabolic rate from measurement of O_2 usage.

A 2-11

(a) **No** The barrier to diffusion is a whole order of magnitude thinner—0.5 to 1 μm.

(b) **No** The surfactant (Type II alveolar epithelial) cells are relatively rounded structures, without cytoplasmic extensions. It is the alveolar epithelial Type I cells which have an attenuated layer of cytoplasm lining the alveoli, and forming tight junctions with cytoplasm of neighbouring similar cells. The surfactant is secreted from the Type II cells into the alveolus, and spreads over the surface of the Type I cell epithelium.

(c) **Yes** One of the effects of gravity is to expand the uppermost, and relatively 'squash' the lowermost air spaces. (But note that the larger expanded alveoli are *ventilated less* than smaller ones, because they are less readily expanded).

(d) **No** Any excess of fluid lost from capillaries in the lungs is drained, as elsewhere in the body, via lymphatics. Fluid seeps through connective tissue between alveoli, to reach the lymphatics which extend to the end of the bronchial tree. Only when a considerable fluid pressure builds up in the interstitial tissue, does oedema extend into the thin alveolar-capillary interface, and ultimately break through the alveolar epithelium into the air-spaces.

(e) **No** There are no cilia between alveoli and respiratory bronchioles. Macrophages must migrate to reach the 'staircase' of cilia which starts at the proximal end of respiratory bronchioles.

A 2-12

(a) **Yes** A decrease in alveolar ventilation below the

level required to keep PCO_2 normal, will necessarily cause a fall in O_2 also.

(b) **No** Low haemoglobin concentration results in there being less oxygen carried per ml of blood, but the PO_2 will be normal if the lungs are normal.

(c) **No** As for (b), CO poisoning causes the 'anaemic' type of hypoxia, in that CO occupies oxygen-carrying sites in Hb, thus putting it out of action for O_2 transport. The PO_2 again is normal if the lungs are functioning normally. Dissolved O_2 is normal.

(d) **Yes** At high altitude the total barometric pressure is low, so 20% oxygen in inspired air represents progressively less than 160 mmHg partial pressure at increasing height above sea level. As inspired PO_2 decreases, alveolar PO_2 decreases, likewise arterial PO_2.

(e) **Yes** A greater than normal divergence of ventilation-perfusion ratios in the lungs results in some mixed venous blood being relatively poorly oxygenated in its passage through underventilated regions of the lungs. The smaller amount flowing past overventilated regions does not make up for this, and arterial hypoxaemia is the result.

A 2-13

(a) **Yes** If PCO_2 is lower than normal, there is by definition a state of hyperventilation. The normal PCO_2 is around 5.2 kPa (40 mmHg).

(b) **No** This is the expected increase in alveolar oxygen partial pressure when alveolar CO_2 pressure is this much decreased.

(c) **Yes** In normal people, the arterial PCO_2 is within 0.4 kPa (3 mmHg) of the alveolar PCO_2.

(d) **No** In normal people, the arterial PO_2 will be not more than 1.4 kPa (10 mmHg) less than alveolar; the rise in PaO_2 during hyperventilation does not, however, appreciably increase the Hb saturation or content of O_2 in arterial blood.

(e) **Yes** A decrease of PCO_2 depletes plasma bicarbonate and must cause an alkalaemia in accordance with the interrelation between pH, bicarbonate and PCO_2, as defined in the Henderson-Hesselbalch equation. Even if the hyperventilation has continued for a long time, and pH has been restored to near normal by other compensatory mechanisms, there is still by definition a respiratory alkalosis when PCO_2 is low.

A 2-14

(a) **No** There is a small transient depression, but as this leads to a rise in PCO_2, it is promptly corrected and ventilation would be as before in 1 min.

(b) **Yes** See (a).

(c) **No** This would mean that ventilation had been stimulated, since PCO_2 depends on the balance between metabolic production and alveolar ventilation, not on the inspired O_2 concentration.

(d) **Yes** After 5 minutes virtually all the nitrogen would be washed out so O_2, CO_2 and water vapour must add up to 1 atmosphere.

(e) **Yes** The break point in breath-holding depends on the combined stimulus of increasing PCO_2 and decreasing PO_2. After breathing oxygen, the rate of rise of PCO_2 would be unchanged, but the hypoxic enhancement of the stimulus is absent.

A 2-15

(a) **No** In healthy people, arterial PCO_2 remains normal in exercise.

(b) **No** They are stimulated only by the types of hypoxia which reduce the arterial oxygen tension (hypoxaemia, 'hypoxic hypoxia').

(c) **No** Afferents from stretch receptors, irritant receptors and 'J' receptors can certainly take part in the control of breathing. It is known however, that the pattern of breathing at rest is not dependent on intact vagus nerves in adult man, as opposed to animals and infants.

(d) **Yes** Although higher centres normally influence breathing, it can continue when only the brain stem is intact and functioning.

(e) **Yes** Voluntary pathways 'bypass' the medullary centres and activate spinal motoneurones.

A 2-16

(a) **Yes** Surface tension forces tend to create a 'suction' drawing fluid out of capillaries. Surfactant counteracts this.

(b) **Yes** There is some net loss of fluid and a steady small lymph flow.

(c) **No** Fluid escapes into the interstitial corners between alveoli; it only bursts through into alveoli when the pressure has risen and increased lymph flow cannot keep up with it.

(d) **Yes** As in any vascular bed.

(e) **No** It is the intravascular pressure which is the driving force, not flow.

A 2-17

(a) **Yes** Mixed expired $CO_2\% \times$ total expired volume/minute $= CO_2$ output. End-expired $CO_2\% \times$ alveolar expired volume/min $= CO_2$ output. So alveolar ventilation $\times 5 = 4 \times 3.5$ giving alveolar ventilation $= 2.8$ l/min.

(b) **Yes** See (a) for alveolar ventilation. $(4 - 2.8)/4$ gives dead space ventilation/total ventilation which is the same as dead space/tidal volume.

(c) **No** Oxygen consumption is 190 ml/min. Calculated CO_2 output (see (a)) $= 140$ ml/min. $RQ = CO_2$ output/O_2 consumption $= 140/190 = 0.73$ which is too low to represent carbohydrate metabolism.

(d) **No** Arterial O_2 content is 200 ml per litre. Consumption is 190 ml/min, from a blood flow (cardiac output) of 4.5 l/min. 190 ml must be removed from 4.5 l, or 42 ml from each litre. So the mixed venous oxygen content will be 200 minus 42 = 158 ml/l.

(e) **No** With end-expired CO_2 5%, this means arterial PCO_2 is slightly lower than normal—if anything there would be a respiratory alkalosis.

A 2-18

(a) **No** Tidal volume at rest $= 4/10 = 0.4$ litre; in exercise $30/30 = 1$ litre.

(b) **No** At maximal working capacity, the heart rate is at a maximum; everyone's maximal heart rate is around 180 to 190; so this subject would be able to work harder.

(c) **Yes** CO_2 output is given by ventilation $\times$ mixed expired $CO_2\%$.

CO_2 output $= RQ \times O_2$ consumption
$= 0.9 \times 1.35$
$=$ ventilation $\times$ mixed expired $CO_2\%$.

Therefore mixed expired CO_2
$= (0.9 \times 1.35)/30 = 4.05\%$.

(d) **Yes** 200 ml/litre in arterial blood and 100 ml/litre in mixed venous blood, means that 100 ml × 13.5 = 1350 ml of oxygen has been removed from the blood per min.

(e) **No** In healthy people, the arterial PCO_2 remains normal in exercise, or may decrease slightly. This value is high.

A 2-19

(a) **No** The blood volume is normal for a 70-kg subject.

(b) **Yes** The arterial PCO_2 is raised, which is by definition a respiratory acidosis.

(c) **Yes** The right atrial pressure is raised.

(d) **Yes** If the PCO_2 is higher than normal, the arterial PO_2 is inevitably lower than normal, when the inspired gas is air.

(e) **No** In respiratory acidosis, due to CO_2 retention, the plasma bicarbonate will be raised, and also there is increased reabsorption of bicarbonate in the kidneys.

A 2-20

(a) **Yes** From the Fick principle, cardiac output = oxygen consumption divided by the a-v difference for oxygen content.

(b) **No** Arteriovenous difference is 50 ml/litre, or 5 ml/dl. Arterial oxygen content is (1.34 × 13.5) ml/dl in Hb and 0.3 ml/dl in solution giving 18.4 ml/dl. So mixed venous is lower than 15 viz 13.39 ml/dl.

(c) **Yes** This follows from the equation relating the three variables (see (a)).

(d) **Yes** Oxygen in solution is simply proportional to the PO_2. So it has increased sixfold, from 0.3 to 1.8 ml/dl.

(e) **No** In normal lungs, the blood will virtually equilibrate with the alveolar gas so arterial PO_2 will rise similarly to alveolar.

A 2-21

(a) **Yes**

(b) **No** Pulmonary artery pressure is normally about 25 mmHg systolic.

(c) **Yes** If there is 200 ml/litre of oxygen in the blood, this implies a Hb concentration of 15 g/dl if it is fully saturated, and more if it is not fully saturated. (This might be a chronically hypoxic patient, in whom there is a high haematocrit, and high Hb concentration).

(d) **No** From the Fick equation, cardiac output = O_2 consumption/arteriovenous O_2 difference = (0.30)/(200 − 140) litre per min/ml per litre = 5 l/min.

(e) **No** This is a normal to high resting oxygen consumption.

A 2-22

(a) **Yes** Hypoxia causes pulmonary vasoconstriction.

(b) **No** There is an increase in the red cell mass, and in haematocrit, causing an increase in viscosity.

(c) **Yes** This follows from the stimulation of ventilation by hypoxia.

(d) **No** Hypocapnia causes alkalaemia: a decreased rate of bicarbonate reabsorption is the compensatory response.

(e) **Yes**

A 2-23

(a) **No** Carotid body, not sinus.

(b) **Yes**

(c) **Yes**

(d) **Yes** A shift to the right allows more ready offloading of oxygen in the tissues and enables maintenance of a higher tissue PO_2. This competes with shift to the left due to hypocapnia.

(e) **No** There is hypocapnia because ventilation is stimulated. Only in hypoxia so severe as to be fatal is the tissue oxygen consumption diminished.

A 2-24

(a) **Yes** Inspired oxygen is twice normal; because alveolar PCO_2 does not change, the alveolar oxygen percentage is higher than at one atmosphere. So the alveolar and arterial partial pressures are more than twice normal.

(b) **No** Arterial PCO_2 is determined by the control of ventilation: it is kept at the normal value; therefore the *percentage* CO_2 will be smaller because the total pressure is higher, and the alveolar partial pressure unchanged.

(c) **Yes** See (b).

(d) **Yes** The arterial PN_2 is related to alveolar PN_2 and so will be more than twice normal.

(e) No The pressure of the inspired gas and of the surroundings is the same, so lung volume will be normal.

A 2-25

(a) Yes

(b) Yes Inspired oxygen has the same percentage value, but is at about 2 atmospheres pressure, so the partial pressure is about twice that at sea level.

(c) No This is an ambient pressure of about 4 atmospheres. Oxygen toxicity would occur.

(d) No PCO_2 is regulated by the usual mechanisms for the control of ventilation.

(e) Yes If there is time during exposure for equilibration of tissues with the raised nitrogen partial pressure, there will be a greater amount dissolved when there is a greater amount of fat. Therefore it takes longer to wash out during gradual decompression, if bubble formation is to be avoided.

A 2-26

(a) Yes The ventilation increases as metabolic activity increases so that $PaCO_2$ is kept normal.

(b) Yes The point d would reflect a new steady state at higher ventilation, but the same CO_2 output as before.

(c) No This could not be so because the ventilation is unchanged. (This arrow could represent an increase in metabolic CO_2 production or addition of CO_2 to the inspired gas, in a patient on constant controlled ventilation.)

(d) Yes A larger subject would have a greater ventilation at any given PCO_2; his CO_2 output at rest would be greater. His resting condition would be represented at b. (The curve could equally refer to the same subject as the solid curve, but at an increased metabolic rate.)

(e) Yes The alveolar ventilation shown is about 4 l/min, PCO_2 just over 5 kPa; PCO_2 in kPa is approximately the same as % CO_2 in alveolar gas; 5% of 4 l/min = 200 ml/min.

Section 3 Blood, Heart and Circulation

A 3-01

(a) **Yes**

(b) **Yes** Mean corpuscular volume (MCV) is the PCV divided by the erythrocyte count.

(c) **Yes** The cells occupy 0.45 l/l so that plasma occupies 0.55 l/l; this is 55%.

(d) **Yes** Mean corpuscular Hb concentration is Hb concentration divided by PCV which gives a value of about 33 g/dl.

(e) **Yes** Macrocytosis is indicated by the mean corpuscular volume, which is normally not less than about 80 fl. Anaemia is indicated by a haemoglobin concentration of 12 g/dl or less.

A 3-02

(a) **No** Microcytosis means that the average volume of the erythrocytes is low. Consequently the mean corpuscular volume (MCV) is the measurement needed to make the diagnosis; this requires packed cell volume (PCV) as well as the red cell count.

(b) **No** Anaemia means a reduction in the concentration of haemoglobin in the blood. This is measured directly by the haemoglobin (g/dl of blood) and is also indicated, less directly, by the red cell count (cells per litre of blood). In anaemia, cells may be small, normal or large.

(c) **Yes** Hypochromia means a reduction in the concentration of haemoglobin in the erythrocytes and is measured directly by the MCHC (given as haemoglobin in g/dl of cells).

(d) **Yes** Leucocytosis means a raised white cell count.

(e) **No** Erythrocyte fragility is measured by suspending samples of blood in solutions containing various concentrations of sodium chloride and noting the osmolarity below which haemolysis occurs. The packed cell volume means the volume of red cells contained in a given volume of blood.

A 3-03

(a) **No** There can never be complete removal of oxygen by the tissues in life, even very active tissues; arm vein blood is a mixture from skin and muscle, so its desaturation will depend on muscular activity, and on skin blood flow.

(b) Yes Bilirubin is formed from haem in the reticulo-endothelial system; it is then secreted in the bile; the iron removed from haem is retained for RBC formation.

(c) Yes

(d) Yes This is the essence of the Sahli method.

(e) Yes If the iron is oxidized to the ferric form, methaemoglobin is formed.

A 3-04

(a) No This allows the relative numbers of different cell types to be counted.

(b) No Some platelets are lost or destroyed in the preparation of the film.

(c) No Reticulocytes are precursors of mature red cells.

(d) Yes This provides a compensatory mechanism for the production of more cells to carry more O_2.

(e) No They live for about 120 days.

A 3-05

(a) No Group O is the universal donor group because the red cells possess no A nor B antigen.

(b) Yes

(c) Yes

(d) Yes Due to excessive haemolysis.

(e) Yes Agglutination occurs in the cold and in the absence of complement.

A 3-06

(a) No If father is heterozygous Rh/rh, then the child has a 50% chance of being Rh negative and hence not at risk.

(b) Yes Previous sensitization of the mother by a Rh-positive baby means that the second Rh-positive child is more at risk than the first.

(c) No It requires repeated exposure of Rh-negative subjects to the Rh antigen for antibodies to be generated.

(d) Yes It is only the incomplete antibodies that cross the placental barrier although maternal antibodies are both complete and incomplete.

(e) No This would exacerbate the haemolysis of the baby's cells by maternal antibodies.

A 3-07

(a) No pH falls as CO_2 is added.

(b) Yes CO_2 diffuses into plasma and RBC; bicarbonate is formed faster in RBC because of carbonic anhydrase; bicarbonate moves out of RBC.

(c) **No** Chloride ions pass into cells, electrically balancing the outward movement of bicarbonate.

(d) **Yes** The rise in CO_2 concentration shifts the curve to the right (Bohr effect).

(e) **Yes** Although the volume of blood per minute flowing through the whole systemic capillary bed must be the same as that in the aorta, flow is slower in capillaries, because total cross-section is greater.

A 3-08

(a) **Yes**

(b) **No** The excitation passes through the septum to the apex then through the major part of the ventricular myocardium finally to invade the posterobasal part of the left ventricle and the pulmonary conus.

(c) **No** The action potential in the ventricular muscle fibre lasts about 300 msec; that in skeletal muscle lasts about 3 msec.

(d) **No** The absolute refractory period of the ventricle starts with the QRS complex of the ECG.

(e) **Yes**

A 3-09

(a) **Yes**

(b) **Yes** This reflects the fact that the duration of diastole changes more than that of systole.

(c) **No** Because the heart rate obviously varies with respiration; the denervated heart does not exhibit sinus arrhythmia. Also, the heart rate would be about 100/min in a denervated heart.

(d) **No** The P–R interval is less than 0.2 sec which is the upper limit of normal.

(e) **No** The two vectors are approximately at right angles to each other.

A 3-10

(a) **Yes**

(b) **Yes** Whilst repolarization of the atria cannot be identified in the ECG because it is obscured by the ventricular complex, it coincides with the Q wave.

(c) **Yes** 0.2 sec is the upper limit of normal.

(d) **Yes**

(e) **Yes** This represents the 'plateau' of the ventricular action potential.

A 3-11

(a) **Yes**

(b) **Yes** For this reason hyperkalaemia can be fatal.

(c) **No** Consult diagrams of action potential in ventricular muscle: the ECG deflections are maximal when there is a change in potential recorded from muscle fibre: ECG is isoelectric during plateau of action potential.

(d) **Yes** This is the effect of acetylcholine—the rate of change of the pacemaker potential is slowed, so that it takes longer to reach threshold.

(e) **No** It is due to an increase in sodium permeability followed by a slower increase in calcium permeability.

A 3-12

(a) **Yes**

(b) **Yes**

(c) **Yes** It is closed between T and R approx.

(d) **No** The first sound, coincident with closure of the atrioventricular valves, occurs just after the QRS complex.

(e) **No** The second sound, coincident with closure of the aortic and pulmonary valves, occurs soon after the T wave.

A 3-13

(a) **No** The activity is being conducted through the atria at this time.

(b) **No** Systole continues until repolarization, at the T wave.

(c) **No** The other way round: if both are blocked, the heart rate rises.

(d) **No** It must rise from around zero to the aortic diastolic pressure, which means a rise of at least 60 mmHg.

(e) **No** The rate of emptying reaches a high peak immediately after opening of the aortic valve and declines very rapidly from that peak.

A 3-14

(a) **Yes** T-P represents most of diastole which occupies a longer interval than the remainder of the cycle at rest, but would be relatively shorter if heart rate were as high as 120.

(b) **No** The ventricles fill most rapidly in early diastole.

(c) **Yes** During repolarization (T wave) potassium permeability is high—potassium is entering cells, restoring membrane potential.

(d) Yes Atrial contraction is initiated by the electrical activity reflected in the P wave.

(e) No Ventricular systole starts with the QRS complex, but continues longer, until the repolarization, reflected by the T wave.

A 3-15

(a) Yes

(b) No The right ventricular pressure varies between about zero during ventricular diastole and about 25 mmHg at the peak of ventricular systole.

(c) No The mean jugular venous pressure is between + 2 and − 2 cm of blood or saline.

(d) Yes The v wave of the acv complex is due to the venous filling of the right atrium which precedes the opening of the atrioventricular valves at the end of ventricular relaxation.

(e) No The first heart sound occurs early in ventricular systole: at this time, the atria have finished active contraction.

A 3-16

(a) No There may be small discrepancies beat by beat, but over several beats, or over several seconds, the output of each ventricle must be the same.

(b) Yes This pressure difference is what causes it to open.

(c) No The stroke volume may well be twice the resting value, but this is partly or mostly because of extra emptying, rather than extra filling.

(d) No The pulmonary valve opens when the right ventricular pressure reaches the falling (diastolic) pressure in the pulmonary artery, which is 8 to 10 mmHg.

(e) No The left ventricular (and right ventricular) pressure falls to near zero in diastole.

A 3-17

(a) No The right ventricular and pulmonary artery pressures rise to only 20 to 25 mmHg during systole.

(b) Yes The pulmonary valve will have closed around a, so the ventricle is relaxing without filling (i.e. isovolumetrically) until the pressure falls below that in the right atrium.

(c) No The pulmonary valve is closed in this phase of the cardiac cycle (see (b)).

(d) **Yes** This is the filling phase—ventricular pressure is below atrial pressure between b and c, and the valve will not close again until the ventricle starts to contract and raise the pressure above atrial, after c.

(e) **No** The peak of the pressure curve is during ventricular contraction, but later than the QRS complex.

A 3-18

(a) **No** The phase 1 to 2 represents isovolumetric contraction.

(b) **Yes** This represents ventricular filling.

(c) **No** It is around 70 ml (= stroke volume).

(d) **Yes** This represents aortic diastolic pressure.

(e) **No** The shaded area represents the work done by the left ventricle during one cardiac cycle.

A 3-19

(a) **No** It is the ventricular volume during isometric relaxation (end-systolic volume).

(b) **Yes** This implies greater emptying.

(c) **Yes**

(d) **Yes** This curve indicates greater pressure development for a given initial volume.

(e) **No** Atropine blocks vagal effects but there is little vagal innervation of the ventricle. Atropine would therefore be without significant effect on the curve Aa.

A 3-20

(a) **Yes**

(b) **No** Sympathetic stimulation does increase the force of ventricular contraction, but vagal stimulation does not decrease it.

(c) **Yes**

(d) **No** Diastole shortens when heart rate increases. Systole shortens very little, though significantly.

(e) **Yes** The vagal slowing effect is predominant over the sympathetic, so removal of both influences would lead to a higher heart rate.

A 3-21

(a) **No** Resistance is inversely proportional to the fourth power of the diameter.

(b) **Yes**

(c) **No** Due to reflection of the pressure wave from branch points of the arterial tree, the pulse pressure is greater in a medium sized artery than in the aorta.

(d) **Yes**

(e) **No** The pressure in the capillaries will fall in this situation.

A 3-22

(a) **No** Since mean blood pressure is proportional to cardiac output × peripheral resistance, the cardiac output must have risen.

(b) **Yes** Because mean blood pressure equals diastolic pressure plus one-third pulse pressure.

(c) **No** Since cardiac output equals heart rate × stroke volume, the stroke volume must have risen.

(d) **Yes**

(e) **No** The perfusion pressure must have fallen. Since this depends on cardiac output × total peripheral resistance, the fall could be due to any combination of changes in these variables.

A 3-23

(a) **Yes** The mean pressure during the cardiac cycle is nearer to diastolic than systolic, and is approximately diastolic plus one-third pulse pressure.

(b) **Yes** The rise of pressure during ventricular systole is related to the force of contraction, but it will also be modified by the compliance of the arterial tree and the total peripheral resistance.

(c) **No** The ventricle is still ejecting blood into the aorta between b and c, that is, until the dicrotic notch which signifies the point of closure of the aortic valve.

(d) **No** The fall-off of pressure in diastole will be related to both the peripheral resistance and the elastic recoil of the arteries.

(e) **Yes** The pulse pressure is greater but the mean pressure is a little lower (see A 3-21(c)).

A 3-24

(a) **No** A decrease in cardiac output can be compensated by an increase in peripheral resistance, so that arterial BP is unchanged.

(b) **No** In exercise, although the systolic BP increases, the mean BP is usually relatively unaltered. So the considerable increase in cardiac output must be balanced by a comparable decrease in peripheral resistance.

(c) **Yes** See also 3–19.

(d) **No** The main factor is a decrease in the vascular resistance in the muscle itself.

(e) **Yes**

A 3-25

(a) **Yes** Systolic and diastolic, therefore mean, arterial pressure rise in this form of exercise, assisting the inflow of blood against the resistance of the 'squeezed' vessels. The reflex originates from receptors in the muscle.

(b) **Yes** The fall in venous return, hence fall in cardiac output, hence fall in arterial BP, lead to a baroreceptor reflex increase in heart rate.

(c) **Yes** For example, if BP falls due to blood loss, venoconstriction decreases the capacity of the circulation and corrects the right heart filling pressure.

(d) **No** A fall in BP can lead to chemoreceptor stimulus, thus increasing ventilation.

(e) **Yes**

A 3-26

(a) **No** Heart rate becomes slower: venous return increases, cardiac output increases, rising BP stimulates baroreceptors.

(b) **Yes**

(c) **No** Cerebral blood flow remains unchanged: a higher BP would be accompanied by 'autoregulatory' vasoconstriction in the brain.

(d) **Yes** Gravity determines blood flow distribution to a great extent in the lungs. The perfusion pressure to the apices increases.

(e) **No** Constriction of lower limb veins occurs on standing.

A 3-27

(a) **No** There is pressure autoregulation in the brain, maintaining constant flow when BP changes.

(b) **Yes**

(c) **Yes** Although the tubular cells have a high oxygen consumption, the renal blood flow is still relatively much higher and the arteriovenous oxygen difference is small.

(d) **No** Hyperventilation reduces arterial PCO_2, which in turn causes cerebral vasoconstriction and reduced blood flow.

(e) **No** Sympathetic cholinergic and beta-adrenergic effects assist vasodilatation but the main factors are local metabolites.

A 3-28

(a) **Yes** Compensation for a slow, moderate blood loss can be continuous and complete—as blood is lost, there is vasoconstriction in the skin and splanchnic areas, and venoconstriction which reduces the capacity of the circulation; HR is also increased, so that cardiac output and BP can be maintained.

(b) **Yes** See (a).

(c) **Yes** Capillary blood pressure will be reduced, so transudation diminishes, so lymph flow diminishes.

(d) **No** This is a nonsense: if the arterial BP is normal or low, constriction would diminish the blood flow further; cerebral vasodilatation accompanies a fall in arterial blood pressure due to haemorrhage.

(e) **No** There is nothing happening which could increase extracellular osmolality.

A 3-29

(a) **Yes** At 50 kg she will have a blood volume of 3 to 4 litres.

(b) **Yes** 5×10^6 RBC per $mm^3 = 5 \times 10^{12}$ per litre.

(c) **Yes** Compensatory mechanisms, via an increased sympathetic outflow, increase heart rate.

(d) **No** The first compensation for the reduced blood volume is a transfer of fluid from the interstitial compartment, which results from lowered capillary blood pressure; this dilutes the red cell mass, so haematocrit decreases.

(e) **Yes** Iron absorption is varied according to requirement—if the total body iron is diminished by blood loss, more is absorbed for haemopoiesis.

A 3-30

(a) **Yes** Veins distend more, with very little pressure increase, up to the point of fullness, when pressure increases rapidly.

(b) **Yes**

(c) **No** This is true only in certain regions e.g. endocrine glands.

(d) **Yes**

(e) **No** Veins are also sympathetically innervated.

A 3-31

(a) **Yes**

(b) **No** They have tight junctions between the endothelial cells which contribute to blood-brain barrier function.

(c) **Yes**

(d) **No** Capillaries are about the same diameter or smaller: RBC are sometimes distorted in passage through.

(e) **No** They cease to fill when precapillary sphincters constrict.

A 3-32

(a) **No** Young healthy individuals have approximately the same maximal heart rate, however fit or unfit (around 190); the difference is in the severity of exercise at which they reach that rate.

(b) **Yes** This is partly because training improves the possible increase in stroke volume so that cardiac output is greater at a given heart rate; partly also because the work done at a given cardiac output is increased by more efficient oxygen extraction in the working muscles.

(c) **No** Cardiac output can increase by about $\times 5$. Since oxygen supply can increase $\times 15$, greater extraction from the blood must account for $\times 3$.

(d) **No** It decreases as the cardiac output increases.

(e) **No** If it changes at all, the diastolic BP decreases in rhythmic exercise, because of the lowering of peripheral resistance by vasodilatation in the muscles.

A 3-33

(a) **No** Vasodilatation is virtually instantaneous when muscle activity starts.

(b) **No** Total cerebral blood flow does not change: if the mean arterial BP rises, brain blood flow is prevented from rising by cerebral vasoconstriction. (Modern methods of mapping regional cerebral blood flow show a localized increase in those areas of the brain 'switched on' during motor activity in exercise: this is a very small local adjustment of blood flow to meet the needs of increased neuronal metabolism).

(c) **Yes**

(d) **Yes** Capillary blood pressure and surface area are greater than at rest because of opening of more channels; more fluid leaves the bloodstream, and the muscle pump assists lymph flow.

(e) **Yes** (*vide supra*)

A 3-34

(a) **No** Some of the blood flows through the portal vein to the liver; the rest passes via the ductus venosus to the IVC.

(b) **Yes** This is another way of saying that the HbO_2 dissociation curve is further to the left for fetal Hb than for adult Hb.

(c) **Yes**

(d) **Yes** The lungs take up some oxygen from the relatively small pulmonary blood flow. The rest of the blood goes directly from right to left.

(e) **Yes** The uninflated lungs present a greater vascular resistance; only a small fraction of blood returning to the right heart passes through the pulmonary circulation: most flows through the pathways of lower resistance—the foramen ovale and ductus arteriosus.

Section 4 Cell Structure and Function

A 4-01

(a) **Yes**

(b) **Yes** This is associated with protective function: foreign particles and mucus are continually carried upwards.

(c) **Yes** There is no exception to this, from mouth to anus.

(d) **Yes** These allow transmission of excitation through a sheet of muscle fibres.

(e) **Yes** They normally constitute 60 to 70% of all white cells.

A 4-02

(a) **No** Calcium is released from the endoplasmic reticulum.

(b) **No** Oxyntic cells secrete acid.

(c) **Yes**

(d) **Yes** They secrete testosterone.

(e) **No** They are concerned with milk ejection.

A 4-03

(a) No They harbour enzymes and these take part in the final processes of digestion itself.

(b) Yes Active transport of Na^+ out of the cell occurs on other surfaces, thus maintaining low intracellular Na^+ and causing passive movement into the cell from the tubular fluid.

(c) Yes

(d) Yes They are concerned both with synthesis and with reabsorption of colloid.

(e) No The antrum is the part of the stomach where there are *not* any parietal (oxyntic) cells: no acid, and no intrinsic factor originates here.

A 4-04

(a) Yes

(b) No They are lamellar structures, not vesicles.

(c) No They are associated essentially with aerobic metabolism.

(d) No There are none in RBCs which metabolize anaerobically.

(e) No Chromosomes are nuclear material.

A 4-05

(a) No Hypothermia slows down energy-dependent processes.

(b) Yes

(c) Yes

(d) Yes

(e) Yes When cells stop actively adjusting the distribution of ions across the membrane, they swell.

A 4-06

(a) Yes This is the definition of the equilibrium potential.

(b) Yes

(c) No Would be zero if the concentrations on the two sides of the membrane were equal.

(d) No Equilibrium potential is defined as the potential which would be observed assuming the membrane were permeable. A state of impermeability to the ion does not influence the equilibrium potential.

(e) No $E_{Na} = 45$ mV
$E_K = -90$ mV approximately.

A 4-07

(a) Yes

(b) No Glucose is not lipid-soluble.

(c) **Yes**
(d) **Yes**
(e) **No** Lipid-soluble substances enter by dissolving in the membrane.

A 4-08
(a) **No** Osmotic pressure depends on the number of particles per unit volume. 5 g/litre of urea has about 1000 times as many particles as 5 g/litre of Hb.
(b) **No** It is about 6.7 atmospheres (>5000 mmHg).
(c) **No** The cells will lose water by osmosis. They will shrink.
(d) **No** Haemolysis occurs when cells swell in a hypotonic solution.
(e) **Yes**

Section 5 Nervous System and Muscle

A 5-01
(a) **Yes** For instance, the skin of the ear lobe only contains free nerve endings and yet cooling is readily appreciated.
(b) **Yes** This reflects the punctate distribution of nerve endings.
(c) **No** The only nerve fibres in the human which conduct at this speed are Group 1A afferents from muscle spindle receptors.
(d) **Yes**
(e) **No** They lie in the dermis.

A 5-02
(a) **Yes**
(b) **Yes**
(c) **Yes**
(d) **No** Rods are sensitive to the green part of the spectrum. If one wears red spectacles in bright light, rods are not bleached; this is a means of dark-adapting the eyes in a bright environment.
(e) **No** Pacinian corpuscles are sensitive to pressure.

A 5-03
(a) **No** This takes the strain off the muscle spindle and reduces its firing.
(b) **Yes** This stretches the spindle receptors.
(c) **Yes** Because the contractile portion is in series with the stretch receptor.

(d) **No** It is striped muscle the fibres of which are thinner than extrafusal muscle fibres.

(e) **Yes**

A 5-04

(a) **Yes**

(b) **Yes**

(c) **No** The high amplitude voltage excursion is an action potential.

(d) **Yes** The first two action potentials occur in rapid succession; the interval between successive impulses increases. This is adaptation.

(e) **No** The action potential peak voltage is constant, consistent with its 'all or none' nature.

A 5-05

(a) **Yes** Because the fovea, on the visual axis, is devoid of rods; these are the elements responsible for vision in dim light.

(b) **No** Because in this situation cones are not excited.

(c) **Yes** The writing is closer than the subject's near point. The pinhole results in only the central part of the lens being used for focussing and this improves the resolving power of the eye.

(d) **No** The pathways for the pupillary reflexes pass from the optic tract to the midbrain without traversing the cortex.

(e) **No** The opposite occurs.

A 5-06

(a) **Yes** The rectus externus muscle is supplied by the 6th cranial nerve and paralysis of this muscle results in inward rotation of the eye ball.

(b) **No** These movements are jerky; they are known as saccades.

(c) **Yes** Such tracking movements are smooth if the movement being followed is smooth.

(d) **Yes** The placement of the lateral canal allows detection of rotation of the head in the horizontal plane and hence the compensatory eye movements are lateral movements.

(e) **Yes** For instance, optokinetic nystagmus.

A 5-07

(a) **Yes** The Maddox rod test shows that blurring of the visual input to one eye results in deviation of the visual axes.

(b) **No** The factor is around 10,000.

(c) **No** Dark adaptation is principally due to retinal factors.

(d) **No** The anterior surface of the cornea provides two-thirds of the refractory power of the eye.

(e) **No** This so reduces the refractory power of the eye that even an object at a great distance cannot be focussed on the retina.

A 5-08

(a) **Yes** The right half of each retina is denervated causing blindness in the left half of the visual field of both eyes i.e. a left homonymous hemianopia. 'Homonymous' means corresponding parts of the visual field in both eyes. 'Hemianopia' means that half of the visual field is lost.

(b) **Yes** The inner half of each retina is denervated causing blindness in both temporal visual fields. 'Heteronymous' means different parts of the visual fields in the two eyes.

(c) **Yes** The efferent limb of the arc is via the 3rd nerve.

(d) **Yes** The pathway for the direct light reflex is via the retinotectal projection.

(e) **No** This reflex depends on pathways from the oculomotor nuclei to the Edinger-Westphal nucleus and thence via III to the pupil; it does not depend on the retinotectal projection.
Note: A lesion at D causes loss of light reflexes but spares the accommodation–convergence reflex, a condition known as the Argyll– Robertson pupil.

A 5-09

(a) **Yes**

(b) **No** Contraction of the ciliary muscle reduces the tension on the suspensory ligaments of the lens which gets thicker.

(c) **Yes** Multiple-unit smooth muscle is smooth muscle in which the muscle units can be separately controlled by the motor nerves.

(d) **No** The ciliary muscle is innervated by the parasympathetic nervous system and so is not contracted by adrenaline.

(e) **No** It runs with the oculomotor nerve.

A 5-10

(a) **Yes** Since he has 1 dioptre of hypermetropia and 2 dioptres of accommodation, when fully accommodated his eye has strength 1 dioptre; it is therefore focussed 1 m ahead which is by definition his near point.

(b) **Yes** The 'far point' is by definition the point from which rays are brought to a focus on the retina by the relaxed (unaccommodated) eye.

(c) **Yes** He exerts 1 dioptre of accommodation to focus on distant objects.

(d) **No** If accommodation is paralysed, the refractive power is too weak to bring any real object to a focus on the retina.

(e) **Yes** With a + 3D spectacle lens, the power of his eye is 2 dioptres. So the eye will be focussed on objects 50 cm away.

A 5-11

(a) **No** Near point = 0.5 metre in front of his cornea.

(b) **No** The definition of an emmetrope is that his far point is at infinity.

(c) **Yes**

(d) **No** If his accommodation is paralysed, his eyes are focussed on infinity and so objects 1 metre ahead are not seen clearly.

(e) **Yes** The power of the eye plus spectacle lens is then 2 dioptres so it is focussed 0.5 metre = 50 cm ahead of him.

A 5-12

(a) **Yes**

(b) **Yes**

(c) **No** There is a synapse on the pathway.

(d) **No** It leaves at the level of the midbrain.

(e) **No** The receptors are muscarinic.

A 5-13

(a) **Yes** The perception of visual images requires the suppression of the image during the fast phase of nystagmus just as during other saccadic eye movement.

(b) **No** It is a saccadic movement.

(c) **Yes**

(d) **No** The slow phase of nystagmus is in a direction such as to stabilize the visual image; the fast phase occurs to bring the visual axis rapidly back towards the midline when the axis has deviated maximally during the slow phase.

(e) **No** Nystagmus occurs, for instance, if the eyes are closed and the subject is rotated on a Barany chair. This can be detected by electrodes placed to pick up the retinocorneal potential.

A 5-14

(a) **Yes** The plane of these canals is closest to the plane of rotation.

(b) **Yes** Such a movement of the eyes tends to stabilize the visual image.

(c) **No** The frictional forces in the semicircular canals are such that, after half a minute of rotation at a constant angular velocity, there is no movement of the endolymph relative to the walls of the canals and hence no sense of rotation.

(d) **Yes** The inertia of the endolymph gives the subject the impression of a rotation in the opposite sense from the rotation which has just stopped. Hence the appropriate direction for the slow component of the nystagmus is in the same direction as the previous rotation.

(e) **No** He tends to fall sideways because the world seems to be rotating from left to right or right to left.

A 5-15

(a) **No** It takes 20 sec or more for eye movements to be generated.

(b) **No** It is the cooling of the endolymph and the resulting convection currents which stimulate the receptors in the semicircular canals.

(c) **Yes** The cold endolymph is dense and so tends to fall. This is the same movement as produced by a rotation of the nose away from the side of injection. To stabilize the visual image, this would require a lateral deviation of the gaze towards the side of the injection.

(d) **No** This indicates damage to higher centres.

(e) **Yes** Caloric testing yields eye movements in an unconscious patient whose brain stem is functioning.

A 5-16

(a) **Yes**

(b) **Yes**

(c) **Yes** These are activated as the head moves on the trunk and initiates neck reflexes.

(d) **Yes** These are sensitive to linear acceleration.

(e) **Yes** These signal rotational movement.

A 5-17

(a) **Yes** This is a commonly used test for normal auditory function.

(b) **Yes**
(c) **No** The differential sensitivity to frequency is the percentage change in frequency which can just be detected. It is fairly constant over a wide range of frequencies.
(d) **Yes**
(e) **No** The receptors in the semicircular canals are sensitive to angular movements of the head and not to linear acceleration.

A 5-18
(a) **Yes**
(b) **Yes** The auditory system is sensitive to a phase difference of this magnitude.
(c) **No** The intensity threshold is least at about 1 kHz and rises steeply for frequencies below or above this.
(d) **No** The just perceptible difference in intensity rises with the intensity level at which it is tested.
(e) **Yes** The difference between the two is at least threefold in most people.

A 5-19
(a) **No** It is the force that is amplified; the amplitude of movement is reduced.
(b) **No** High frequencies are detected by receptors close to the oval window.
(c) **Yes**
(d) **No** Contraction of the stapedius muscle reduces the effectiveness of mechanical transmission through the ossicles.
(e) **Yes** The energy of the sound waves is conducted directly to the receptors through the bone.

A 5-20
(a) **No** The orientation of the head in space alters and this changes the pattern of stimulation of the otolith receptors.
(b) **No** The forelimbs extend.
(c) **Yes** For the neck reflexes, the change in tonus in limb musculature is such as to push the trunk back into line with the head.
(d) **No** The two tend to cancel. This is probably why a human or other animal is free to move the head on the trunk without initiating inappropriate postural reflexes.

(e) Yes This is a reflex with sensors in the semicircular canals and helps to stabilize the retinal image.

A 5-21

(a) **Yes** The permeability of the nerve membrane is voltage dependent; depolarization causes sodium gates to open.

(b) **Yes** Because the gates which open are specific for sodium ions.

(c) **No** The permeability to potassium increases after the peak of the action potential.

(d) **No** The opening of ionic gates constitutes a reduction in membrane resistance.

(e) **No** Acetylcholine release is not an essential step in propagation of the action potential.

A 5-22

(a) **Yes**

(b) **Yes**

(c) **No** It is less than 0.5 msec.

(d) **No** The resting potential is usually within 20 mV of the chloride equilibrium potential; the voltage of the peak of the action potential is within 20 mV of the sodium equilibrium potential.

(e) **No** The downstroke of the action potential is due largely to passive efflux of potassium ions down their electrochemical gradient.

A 5-23

(a) **No** The conduction velocity is independent of the stimulus strength.

(b) **Yes**

(c) **No** The action potential so elicited would travel in both directions.

(d) **No** 15 to 70 m/sec would be an appropriate range.

(e) **No** The smaller the diameter of an axon, the slower it conducts.

A 5-24

(a) **Yes** These are the Group 1A afferent fibres.

(b) **No** The efferent fibres are smaller than the Group 1A afferents.

(c) **No** The larger nerve fibres are preferentially stimulated.

(d) **Yes**

(e) **No** The M and H responses are both generated by extrafusal muscle fibres. The M is the response to direct stimulation of efferent nerve fibres.

The H is the reflex response to the stimulation of afferent nerve fibres.

A 5-25

(a) **No** There is a flaccid paralysis as in a 'lower motoneurone lesion'.

(b) **Yes**

(c) **No** It will be warmer, due to interruption of the sympathetic vasoconstrictor nerve fibres.

(d) **No** The sweat glands are also denervated.

(e) **Yes**

A 5-26

(a) **Yes**

(b) **Yes** Since there is little movement of water across the cell membrane as a muscle contracts, the muscle volume is essentially constant, so shortening must be accompanied by thickening.

(c) **No** This figure is about 3 times too great.

(d) **No** The force of contraction is inversely related to the rate of shortening.

(e) **No** Each muscle fibre is controlled by a single motor nerve.

A 5-27

(a) **No** This is a property only of skeletal muscle.

(b) **No** Smooth muscle has no striations.

(c) **Yes**

(d) **Yes**

(e) **Yes**

A 5-28

(a) **Yes**

(b) **No** In large muscles subserving strong coarse movements (e.g. gastrocnemius), a motor unit comprises several hundred muscle fibres. In small muscles used for finely controlled movements, (e.g. the extrinsic ocular muscles), the motor unit may comprise about 4 muscle fibres.

(c) **Yes**

(d) **No** Asynchrony in activity in different motor units produces a smooth movement even although each motor unit alone may contribute an unfused contraction.

(e) **Yes**

A 5-29

(a) **Yes**

(b) **Yes** In this respect it is unlike striated muscle in which tension falls off rapidly with reduction in initial length.
(c) **Yes** For instance circulating catecholamines cause the smooth muscle sphincters of the bowel to contract.
(d) **Yes**
(e) **Yes** Many smooth muscles have a double innervation, sympathetic and parasympathetic the effects of which are often in opposite directions.

A 5-30
(a) **Yes**
(b) **Yes** It causes a depolarization block of neuromuscular transmission of short duration.
(c) **Yes**
(d) **Yes** Physostigmine reverses the blockade of curare in this way.
(e) **Yes** Packets of acetylcholine are released spontaneously from the nerve terminals and produce miniature end plate potentials.

A 5-31
(a) **No** Such a change in membrane potential would be inhibitory.
(b) **Yes** For example in the tendon jerk reflex. IPSPs are generated on motoneurones to antagonists of the quadriceps via the orthodromic inhibitory pathway.
(c) **Yes** This is an important mechanism for integration in the central nervous system.
(d) **No** An EPSP is an opening of gates which allow all inorganic ions to pass; the action potential gates are specific for sodium ions.
(e) **Yes**

A 5-32
(a) **No** Probably glycine, certainly not acetylcholine.
(b) **Yes**
(c) **No** They are produced by ions, principally potassium, flowing down their electrochemical gradients.
(d) **No** Renshaw cell feedback is from neighbouring motoneurones also.
(e) **No** Different interneurones are involved.

A 5-33
(a) **No** Probably glycine is the transmitter; glutamate is an excitatory transmitter.

(b) No The IPSP is hyperpolarizing.
(c) No In the cord, inhibition is always mediated via interneurones.
(d) No Being postsynaptic potentials, they are recorded from the motoneurones themselves.
(e) No Passive stretching of a muscle initiates excitation, not inhibition, of the motoneurones in its own pool.

A 5-34
(a) Yes
(b) No This would be true of postsynaptic potentials.
(c) Yes
(d) No The action potentials invade the terminal but, being smaller than usual, they cause the release of less transmitter.
(e) No Presynaptic inhibition is probably due to the opening of channels permeable principally to chloride ions whereas postsynaptic inhibition is due to channels specifically permeable to potassium.

A 5-35
(a) Yes
(b) No The transmitter is probably glutamate, certainly not acetylcholine.
(c) No The membrane will be clamped at the chloride potential and excitatory synaptic drive will consequently be less effective.
(d) Yes Strychnine probably acts by blocking the inhibitory transmittor glycine.
(e) Yes

A 5-36
(a) No It usually innervates a group of muscle fibres called the motor unit.
(b) Yes
(c) Yes
(d) Yes
(e) Yes

A 5-37
(a) No 30 msec would be a correct answer.
(b) No The shortest reaction time is about 100 msec.
(c) Yes Tendon jerks are the only monosynaptic reflexes in the body.

(d) **No** The orthodromic inhibitory pathway is disynaptic.
(e) **Yes**

A 5-38
(a) **No** It consists of first-order neurones.
(b) **No** It carries central afferents subserving fine tactile discrimination, pressure and vibration.
(c) **Yes**
(d) **Yes**
(e) **Yes** The dorsal column carries fibres conveying proprioceptive information and is an important input pathway to the cerebellum.

A 5-39
(a) **Yes**
(b) **Yes**
(c) **No** The lateral spinothalamic tract is crossed.
(d) **Yes**
(e) **No** The cerebellum receives mainly proprioceptive input.

A 5-40
(a) **No** Like somatic spinal reflexes, autonomic spinal reflexes are paralysed at this stage.
(b) **No** There is a flaccid paralysis below the level of the transection.
(c) **Yes**
(d) **No** Consciousness is not impaired.
(e) **No** The blood pressure is usually depressed, both because the influence from the vasomotor centre has been disconnected and because of the temporary cessation of all spinal cord activity.

A 5-41
(a) **Yes**
(b) **Yes**
(c) **No** Temperature afferents travel in the lateral spinothalamic tracts, which are crossed.
(d) **Yes**
(e) **No** Crude touch from each leg travels up the cord bilaterally in both the anterior spinothalamic tracts; crude touch sensation is impaired, not lost, on both sides.

A 5-42

(a) **No** Tendon jerks are increased below the level of the transection.

(b) **Yes**

(c) **Yes** Abdominal and cremasteric reflexes depend on the integrity of the cortico-spinal projection.

(d) **No** An extensor plantar response is present.

(e) **Yes** In spinal man, shivering never occurs in musculature below the level of the lesion.

A 5-43

(a) **No** Micturition is a spinal reflex. The subject will have an 'automatic bladder'.

(b) **No** As the bladder fills, the blood pressure rises; this is unlike the intact human in whom the blood pressure is stabilized by reflexes mediated via higher centres.

(c) **No** This occurs as a component of the mass reflex.

(d) **Yes** Raising the subject from the horizontal to the sitting position results in a fall in blood pressure because there can be no reflex adjustment via the brain stem (except a vagally mediated increase in heart rate).

(e) **No** Many subjects sweat profusely, as a result of uncontrolled spinal sympathetic output; this condition may be hard to control.

A 5-44

(a) **Yes**

(b) **No** They end in the cerebellar nuclei.

(c) **Yes**

(d) **No** They terminate on inhibitory synapses.

(e) **Yes**

A 5-45

(a) **No** Tremor at rest is characteristic of lesions of the basal ganglia.

(b) **Yes**

(c) **No** A lesion of the cauda equina abolishes the Babinski response.

(d) **Yes**

(e) **Yes**

A 5-46

(a) **Yes**

(b) **Yes**

(c) **Yes** Involuntary movements are often regarded as the hallmark of disease of the basal ganglia.

(d) **No** Intention tremor is a sign of cerebellar disease.

(e) **Yes**

A 5-47
(a) **Yes**
(b) **Yes**
(c) **No** They lie in the midbrain reticular formation.
(d) **No** In most subjects, the speech areas are in the left hemisphere.
(e) **No** A lesion of Broca's area usually results in a reduction of talking and 'telegram' speech. Excessive talking is characteristic of a lesion of Wernicke's area.

A 5-48
(a) **Yes** This reaction is a spinal reflex.
(b) **No** The pupils will respond to light, because the brain stem is intact.
(c) **No** Not without the cerebral cortex.
(d) **No** The limbs will be spastic.
(e) **Yes**

A 5-49
(a) **No** These are in the carotid sinuses and aortic arch.
(b) **Yes** TSH inhibits TRH-secreting cells.
(c) **No** These are in the medulla.
(d) **No** These are in the peripheral chemoreceptors.
(e) **No** These are in the atria.

A 5-50
(a) **Yes** Increased activity in fibres to sweat glands is accompanied by decreased activity in fibres to the arterioles supplying the superficial plexus of skin blood vessels.
(b) **Yes**
(c) **Yes**
(d) **No** The autonomic innervation is purely sympathetic.
(e) **No** The synapse lies in a paravertebral ganglion or in a collateral ganglion distant from the skin itself.

A 5-51
(a) **Yes** Because of paralysis of the constrictor pupillae muscle.
(b) **Yes** Because of paralysis of the ciliary muscle.
(c) **No** There is no effect on extrinsic ocular muscles.
(d) **Yes** The canal of Schlemm tends to be held open by contraction of the ciliary muscle.

(e) No Everything tends to appear brighter because of the dilated pupil.

A 5-52

(a) Yes

(b) No The dilator pupillae muscle, supplied by the cervical sympathetic, is paralysed.

(c) No Enophthalmos would be correct.

(d) Yes Due to paralysis of the vasoconstrictor nerve supply.

(e) No Lacrimation is under the control of the parasympathetic nervous system.

A 5-53

(a) Yes

(b) Yes

(c) No Their cell bodies lie in the lateral horn of the grey matter of the spinal cord.

(d) Yes

(e) Yes

Section 6 Homeostasis

A 6-01

(a) No In this situation, the cells would shrink.

(b) No Extracellular sodium is much the higher.

(c) No Extracellular chloride is much the higher.

(d) Yes Intracellular potassium is much the higher.

(e) No Intracellular pH is less than extracellular.

A 6-02

(a) Yes Aldosterone tends to restore blood volume, by reabsorption of sodium, which brings water with it osmotically.

(b) No They become more permeable to water (by the action of ADH) which allows greater water reabsorption along the osmotic gradient.

(c) No The amount reabsorbed in the proximal tubules is always the same proportion of the filtered load. The variation is in the distal tubules, by the action of aldosterone.

(d) Yes Increased ECF osmolality stimulates hypothalamic osmoreceptors, leading to secretion of ADH, greater retention of water, hence correction of osmolality.

(e) No Absorption is active and does not depend on a diffusion gradient. Excess is corrected in the kidneys, not by controlling absorption.

A 6-03
(a) No Diffusion is the continuous movement of molecules in all directions. Diffusion from a region of greater to one of lesser concentration implies a net movement in that direction.
(b) No It is about 1/200.
(c) Yes Cl^- concentration is over 100 mmol/l.
(d) No It is the products, not the sums, which must be equal (Donnan equilibrium).
(e) Yes

A 6-04
(a) Yes Increased venous return leads to higher ventricular filling pressure, and hence to an increase in stroke volume.
(b) Yes Dilution of plasma protein enhances movement of fluid out of capillaries, therefore to greater lymph flow.
(c) No Renin secretion, hence angiotensin, hence aldosterone would decrease, leading to a decrease in Na^+ reabsorption.
(d) No Even though arterial blood pressure may rise, cerebral autoregulatory vasoconstriction keeps blood flow steady.
(e) No Most of the fluid remains extracellular because the osmolality is changed very little.

A 6-05
(a) No Water absorption is independent of intake or need.
(b) Yes Any dilution of the blood will lead to transfer of water across capillary walls, by osmosis.
(c) Yes An increase in blood volume can lead to stimulation of atrial receptors and hence to stimulation of release of the appropriate hormones.
(d) No These are chemoreceptors which are stimulated by low PO_2, high PCO_2 and low pH.
(e) No The main means is by altering reabsorption rate in the distal tubules and collecting ducts.

A 6-06
(a) No About 10% is degraded every day.
(b) No MW 69,000 and 156,000 respectively.

(c) **Yes** Its *molar* concentration accounts for the major part of the colloid osmotic pressure.

(d) **Yes**

(e) **Yes** Because of the lowering of colloid osmotic pressure.

A 6-07

(a) **Yes** Normal CSF pH is about 7.32.

(b) **No** Although formation is not only by passive processes, a hypertonic plasma would diminish the rate of extravasation of fluid—water would tend to move from CSF to plasma.

(c) **No** Active processes are involved.

(d) **Yes** CSF contains less protein than even the small amount in other extracellular, extravascular fluids.

(e) **Yes** Absorption is via the arachnoid villi.

A 6-08

(a) **Yes** Probably about 50% of the CSF is formed at the choroid plexuses; it also escapes in an actively controlled manner from all CNS capillaries into the interstitium and subarachnoid space (across the blood–brain barrier).

(b) **Yes** The whole of the cerebral ventricles and subarachnoid space together with the extracellular fluid are without barriers to diffusion; CSF is an extension of the interstitial fluid of the central nervous system.

(c) **Yes** Potassium concentration is regulated at the blood–brain barrier: cerebral extracellular and CSF $[K^+]$ is just over half that in plasma.

(d) **Yes** The chemosensitive areas near the ventral surface of the medulla are stimulated by acidity of CSF.

(e) **No** The blood–brain barrier is relatively impermeable to H^+.

A 6-09

(a) **Yes**

(b) **Yes**

(c) **No** Plasma albumin is not distributed outside the plasma. This ‘label’ would measure plasma volume only.

(d) **Yes** By ‘insensible perspiration’ when not sweating.

(e) **Yes** Absorption of excessive water from the gut leads to some degree of haemodilution, transient because it is rapidly corrected by osmotic movement into the extravascular compartment and by decreased secretion of ADH.

A 6-10

(a) **Yes** In a state of balance, little is lost by other routes, and virtually all is absorbed from the glomerular filtrate.

(b) **Yes** Accelerates repolarization and leads to eventual arrest.

(c) **Yes** Weakens contraction.

(d) **Yes** This may contribute to vasodilatation in active muscle and neural tissue.

(e) **Yes** Aldosterone stimulates distal tubular potassium secretion in the kidney.

A 6-11

(a) **Yes** Variations in the amount of ammonia formed in and secreted by tubular cells allows adjustable excretion of H^+ in the form of ammonium salts in the urine.

(b) **No** The GFR is about 120 ml/min: if only half of this were retained urine flow would be over 3 litres per hour, which is clearly nonsense.

(c) **Yes** Glucose is all reabsorbed under normal circumstances.

(d) **No** Urea is not secreted by tubular cells: it moves by diffusion: some re-enters the blood; some moves out of the nephron and in again, according to differences in gradients and in permeability along the nephron.

(e) **Yes** Bicarbonate is formed from CO_2 entering both from filtrate and from plasma, catalysed by carbonic anhydrase; bicarbonate then diffuses into the blood.

A 6-12

(a) **No** Normally about one-fifth.

(b) **No** About 80% is reabsorbed here.

(c) **Yes**

(d) **No** The tubular fluid leaving the loop of Henle is hypotonic. It is the tubular fluid in the deepest part of the loop and the interstitial fluid around that part, which become hypertonic.

(e) **Yes** There is an osmotic gradient tending to draw water into the interstices from the tubular fluid and thus to reabsorb it. The more permeable the tubule to water, the more readily is it reabsorbed.

A 6-13

(a) **Yes** Since electrolytes pass freely through into the glomerular filtrate, the concentrations are for practical purposes the same as in plasma. (There are small differences because of the difference in protein concentration).

(b) **Yes** The haematocrit increases during passage through the glomerulus, because some (about one-fifth) of the plasma volume is lost.

(c) **No** There is an important autoregulatory control mechanism which keeps glomerular capillary pressure, and renal blood flow, near constant during fluctuations of arterial blood pressure.

(d) **Yes** Pressure in other capillaries averages about 20 mmHg, unless there is a higher pressure due to gravity. In the glomeruli, it is 60 to 80 mmHg, providing an effective filtration, sufficient to counteract the opposing osmotic force of 25 mmHg oncotic pressure, and the hydrostatic back pressure from the filtered fluid in the capsule.

(e) **Yes** About one-fifth of the plasma volume flowing is filtered off; about one-fifth of the glucose flowing in the plasma is therefore filtered off with the fluid. This does not materially change the concentration of glucose in the plasma which continues out through the efferent arteriole.

A 6-14

(a) **Yes** The urine flow in a non-diuretic subject is about 1 ml/min.

(b) **No** Urinary concentration of glucose = (urinary excretion of glucose per min)/(urinary excretion of water per min) = (0.18 mmol/min)/(0.002 litre/min) = 90 mmol/litre.

(c) **Yes** Renal clearance of glucose = (U(glucose) × V)/P (glucose) = (0.18 mmol/min)/(18 mmol/litre) = 0.01 litres/min.

(d) **Yes** Tm for glucose = (mass of glucose filtered per min) minus (mass of glucose excreted per min) = (18 × 0.125 mmol/min) – (0.18 mmol/min) = (2.25 – 0.18) mmol/min.

(e) **No** Such an inference could only be made if the data indicated a glucose clearance in excess of the glomerular filtration rate.

A 6-15

(a) **Yes**

(b) **No** The clearance of such a substance is the GFR, about one-fifth plasma flow.

(c) **Yes** Inulin is filtered, but not absorbed or secreted, compare (b).

(d) **No** The final concentration of the fluid depends on the osmotic gradient moving water from distal lumen into papillary interstitial fluid.

(e) **No** Aldosterone secretion must decrease if an excess of ECF is to be corrected, so that less sodium and water are reabsorbed.

A 6-16

(a) **Yes**

(b) **No** The substance may also be reabsorbed. This is true, for instance, of potassium.

(c) **Yes** It is usually zero.

(d) **Yes** Glucose will appear in the urine: the glucose clearance will rise above zero.

(e) **Yes** Bicarbonate behaves as though it had a Tm (transport maximum). At plasma levels below about 28 mmol.l^{-1}, bicarbonate is almost completely reabsorbed. Above this plasma level, it spills over into the urine, with a consequent rise in its clearance.

A 6-17

(a) **Yes** Urine volume × urine concentration = excretion rate.

(b) **Yes** This is the formula for, and the definition of, 'plasma clearance'.

(c) **No** The clearance for a substance which is only filtered, gives the glomerular filtration rate, which is about one-fifth of the plasma flow.

(d) **Yes** Insulin is filtered, but not absorbed or secreted, see (c).

(e) **Yes** This means that 'more plasma is cleared' of substance A than of substance B. B is not absorbed, so clearance of B must be equal to or greater than the GFR. So if clearance of A is greater A must be secreted in addition to the amount of the GFR.

A 6-18

(a) **Yes** This is arithmetic: 26 000 − 25 850 = 150; 5% of 26 000 = 1300.

(b) **No** Likewise, 900 − 900 + 100 = 100. 5% of 900 = 45, so more than twice 5% of the filtered load is in the urine.

(c) **No** Urinary loss of K^+ is 100, and of Na^+ is 150, mmol/day.

(d) **Yes** Nearer 30-fold.

(e) **No** The principal anion is Cl^-.

A 6-19

(a) **Yes**

(b) **No** The descending limb is freely permeable only to water.

(c) **No** It is hypertonic.

(d) **No** Electrolyte (probably primarily chloride) is pumped in the opposite direction.

(e) **Yes**

A 6-20

(a) **Yes** The countercurrent mechanism brings this about.

(b) **No** It is necessary for the functioning of the countercurrent mechanism that the ascending tubules are relatively impermeable to water: as Cl^- is pumped out and Na^+ follows, the fluid becomes less hypertonic.

(c) **Yes** Because of water reabsorption.

(d) **No** By far the greatest part of the water is reabsorbed in the proximal tubule, although the final amount is further regulated by the degree of ADH activity affecting the distal tubule.

(e) **No** The relevant hormone is ADH.

A 6-21

(a) **No** The hydrogen ion concentration in such a subject is greater in the luminal fluid than in the tubular cell cytoplasm. So hydrogen ions are pumped against this concentration gradient; an energy-requiring pump mechanism is needed.

(b) **No** The minimum is about 4.5.

(c) **Yes**

(d) **No** The reverse is the case.

(e) **No** This is nonsense. Hydrogen and chloride ions carry opposite charges and so would move in the same direction to preserve electroneutrality. Hydrogen ions are secreted largely in exchange for sodium ions.

A 6-22

(a) **Yes** There is a tendency to metabolic alkalosis, so reabsorption of bicarbonate is decreased in vegetarians.

(b) **Yes** The lactate is metabolized so that effectively sodium hydroxide is released into the body. This metabolic alkalosis promotes renal excretion of bicarbonate.

(c) **No** Ammonia is formed from glutamine in the tubule cells.

(d) **Yes**

(e) **Yes** The major part of excess acid is excreted in the form of ammonium ions.

A 6-23

(a) **Yes** Because of loss of hydrochloric acid.

(b) **Yes** Metabolic alkalosis is associated with a high plasma bicarbonate concentration.

(c) **No** Metabolic alkalosis is usually associated with a low potassium concentration.

(d) **No** Chloride is lost in the vomitus.

(e) **Yes** Despite the metabolic alkalosis, the loss of sodium from the body activates renal sodium retaining mechanisms. Preservation of electroneutrality involves increased excretion of hydrogen ions in the urine. Defence of body electrolytes takes precedence over the defence of pH.

A 6-24

(a) **Yes** The normal value is $10^{-7.4}$ mol/litre ($pH = 7.4$ = negative log of hydrogen ion concentration) which is $10^{-3} \times 10^{-4.4}$ or $10^{-7.4}$mmol/litre.

(b) **Yes** This is the approximate amount generated, so must be excreted to remain in balance.

(c) **Yes** An increase in blood acidity stimulates ventilation, probably mainly via the arterial chemoreceptors (the blood–brain barrier is not freely permeable to H^+, so central chemoreceptors are not directly affected).

(d) **No** Ventilation is stimulated, which decreases the PCO_2, tending to correct acidity. (In a *closed* system, an increase in H^+ would increase the PCO_2 but in the body this is immediately lost in the lungs.)

(e) **No** When acidity *decreases*, less H^+ is excreted by the tubular cells into the tubular fluid, and less CO_2 moves into the cells to form bicarbonate. The effective result is less HCO_3 'reabsorption' in the kidney, which tends to correct the ECF change.

A 6-25

(a) **Yes** This is because the concentration of plasma proteins is higher than that of phosphates.

(b) **No** Renal hydrogen ion secretion is limited by the concentration gradient between tubular cell cytoplasm and tubular fluid.

(c) **Yes** This is partly by a mass action effect.

(d) **Yes**

(e) **No** Loss of a water molecule is equivalent to the loss of one hydrogen and one hydroxyl ion, so the net effect on acid-base balance is zero.

A 6-26

(a) **No** pH = 6.1 + log (24/(0.03 × 40)) = 6.1 + log 20 = 6.1 + 1.3 = 7.4

(b) **Yes** By substitution in the equation.

(c) **Yes**

(d) **No** The bicarbonate buffer system buffers acids and bases other than CO_2.

(e) **Yes** The type of reaction which takes place is: $H^+ + Cl^- + Na^+ + HCO_3^- = Na^+ + Cl^- + H_2O + CO_2$. Excess fixed acid plus plasma bicarbonate yields sodium and chloride ions, water and CO_2, the latter being blown off in the lungs.

A 6-27

(a) **No** There is a direct proportionality between these two variables.

(b) **No** $[HCO_3]/[CO_2]$ is approximately 1000.

(c) **No** From the Henderson-Hasselbalch equation, the pH is unaltered.

(d) **Yes** This is the principal way in which renal buffering of acidosis is effected.

(e) **Yes** Each buffer system is most effective at its pK (in this case 6.1) and progressively less effective as the pH moves away from the pK.

A 6-28

(a) **No** Respiratory acidosis, compensated or not, implies a raised arterial PCO_2.

(b) **Yes** 'Fully compensated' means that the pH has been restored to normal by the kidney.

(c) **No** The plasma bicarbonate concentration must be raised.

(d) **Yes** By the Henderson-Hasselbalch equation.

(e) **No** There will be a positive base excess due to addition to the blood of bicarbonate by the kidney.

A 6-29

(a) **Yes** CO_2 is readily diffusible across all membranes. pH change will be greater or less according to the amount of buffer present, but is a decrease everywhere.

(b) **Yes** This is because CO_2 is freely transferable across the blood–brain barrier, not because H^+ is freely transferable.

(c) **Yes** The renal compensation for respiratory alkalosis is diminished effective reabsorption of bicarbonate.

(d) **Yes** Cerebral vessels constrict when $PaCO_2$ is decreased; this leads to faintness/dizziness.

(e) **Yes** The greater the blood flow, the smaller will be the arteriovenous difference for CO_2, so that given a steady rate of tissue CO_2 production, the local rise in PCO_2 will be less. This applies in any tissue, but particularly in the brain where vessels are sensitively dilated by rises in PCO_2.

A 6-30

(a) **No** This is the action of antidiuretic hormone (ADH).

(b) **Yes**

(c) **No** ADH is released from the posterior pituitary; the control is via the direct neural pathway from the hypothalamus. Releasing hormones act on the anterior pituitary.

(d) **Yes** Receptors in the atrium initiate a neuroendocrine reflex which leads to less aldosterone secretion, therefore less retention of sodium and water, therefore a reduction in blood volume.

(e) **Yes** Aldosterone promotes the secretion of potassium, as well as the retention of sodium, in the kidney.

A 6-31

(a) **Yes**

(b) **No** Minimal water intake must balance obligatory fluid loss which is approximately 1 to 1.5 litres:
urine 450 ml/24 hr
faeces 150 ml
evaporation 500 ml
respiration 300 ml.

(c) **Yes**

(d) **No** About one-tenth of the renal blood flow, or one-fifth of the plasma flow is filtered. Renal blood flow is in turn about one-fifth of cardiac output.

(e) **Yes** This is the total osmolarity; not to be confused with colloid osmotic or oncotic pressure.

A 6-32

(a) **Yes**

(b) **Yes**

(c) **Yes**

(d) **No** They are secreted in the female also.

(e) **Yes** Cerebral extracellular fluid includes that to which hypothalamic osmoreceptors are exposed; increase in osmolality stimulates increased secretion of ADH, so that more water is retained in the kidney, and the extracellular compartment is 'diluted' back to normal.

A 6-33

(a) **No** Aldosterone promotes sodium reabsorption.

(b) **Yes** The respiratory alkalaemia due to lowered blood PCO_2 causes a reduction in the concentration of ionized calcium.

(c) **No** By definition, respiratory alkalosis implies a low $PaCO_2$, even though pH may have been corrected back to normal.

(d) **No** Not unless the lungs are abnormal. The oxygen content is low.

(e) **Yes** Lactic acidosis causes a fall in pH.

A 6-34

(a) **Yes** Secretin acts to reduce duodenal acidity by stimulation of pancreatic (alkaline) secretion.

(b) No Opening of sodium channels assists the progress of depolarization—this is positive feedback.
(c) No Secretion of bile salts by the liver into the bile does not reduce the plasma concentration, so this is not negative feedback.
(d) Yes TSH, by stimulating release of thyroid hormone, tends to correct the low plasma level.
(e) No Just before ovulation, the secretion of LH is stimulated by circulating oestrogen; this is positive feedback.

A 6-35
(a) No That is in the pons. For some reason a common mistake.
(b) Yes
(c) Yes
(d) Yes Posterior pituitary is developed from neural tissue; secretions travel along axons to be released from the posterior pituitary.
(e) Yes Hypothalamic cells secrete releasing factors into the local capillaries, veins carry this blood to the anterior pituitary and break up into a second capillary system therein.

A 6-36
(a) No Liver glycogen in the source.
(b) Yes Plasma urea concentration depends on metabolic production; concentration in glomerular filtrate follows plasma concentration; increased filtered load leads to increased excretion.
(c) No The tubular maximum is higher than is required, at normal concentrations, for complete reabsorption.
(d) Yes Rise in plasma Ca^{++} inhibits parathormone release, which enhances deposition.
(e) Yes The liver is the source of plasma proteins.

A 6-37
(a) Yes There are osmoreceptors in the hypothalamus.
(b) No These are stretch receptors in the arterial wall, sensing changes in arterial blood pressure.
(c) Yes Reflex responses are both neural and endocrine, altering venous capacity and blood volume.
(d) No The main control is the negative feedback

effect of glucose concentration on insulin secretion.

(e) **Yes** Renin-secreting cells of the juxtaglomerular apparatus are activated as a result of decreased blood volume or increased sodium load, probably via sensors in the macula densa or in the walls of the glomerular arterioles.

A 6-38

(a) **Yes** This tends to raise temperature by increasing metabolic rate at rest.

(b) **Yes** Low iodine leads to low production of thyroid hormone, which stimulates secretion of TSH.

(c) **No** Insulin is secreted.

(d) **No** Renin leads, via aldosterone, to sodium and water retention.

(e) **Yes** This is a neuro-endocrine positive feedback mechanism.

A 6-39

(a) **No** The hypothalamic releasing factors are distributed through a special portal system, for example.

(b) **No** A limit is reached when receptor sites are fully occupied.

(c) **Yes** Although many hormones act at a membrane, some enter first and act within e.g. aldosterone.

(d) **Yes** For example insulin.

(e) **No** Radioimmunoassay measures the amount of hormone in a sample, after the sample has been treated with a serum containing antibody to the hormone.

A 6-40

(a) **Yes** For example, over 99% of T4 and T3 in the blood is bound.

(b) **Yes**

(c) **No** Many hormones in the free form have a small enough molecular weight to be filtered.

(d) **No** Because TSH secretion depends on TSH releasing factor from the hypothalamus.

(e) **Yes** Because prolactin secretion is ordinarily suppressed by prolactin inhibiting hormone from the hypothalamus.

A 6-41

(a) **Yes** The lack of iodine prevents adequate synthesis of active hormone; low T3 and T4 in the blood stimulates TSH secretion, which causes hypertrophy of the gland and excessive colloid (goitre).

(b) **No** Thyroglobulin is the form stored in the colloid in the thyroid gland.

(c) **Yes** There is less T3 but it is more active than T4.

(d) **No** The concentration of *active* hormone is separately regulated; *total* hormone increases.

(e) **Yes** This leads to a greater heat production.

A 6-42

(a) **Yes**

(b) **No** The secretions are controlled by releasing factors from the hypothalamus which reach the anterior pituitary via the local portal blood system.

(c) **Yes** Via the hypothalamus and the secretion of GH releasing hormone; GH decreases peripheral glucose utilization and promote lipid utilization; hence the brain is protected from shortage.

(d) **No** The equivalent function in the male is in the control of sperm development.

(e) **Yes** Secretion rises during pregnancy.

A 6-43

(a) **Yes** Reabsorption of phosphate by the proximal tubule is depressed.

(b) **Yes** This metabolite of vitamin D promotes calcium reabsorption from the gut.

(c) **No** Parathormone stimulates osteoclast activity.

(d) **Yes**

(e) **Yes**

A 6-44

(a) **Yes** This is the cause of tetany e.g. in excessive hyperventilation.

(b) **Yes** Low calcium stimulates release of parathormone and diminishes release of calcitonin.

(c) **Yes** The combination of low calcium and high phosphate concentration is found in hypoparathyroidism.

(d) **Yes** This interferes with absorption.

(e) **Yes** A primary defect of parathormone output allows calcium to decrease; (on the other hand, low serum calcium for any other reason will stimulate parathormone secretion).

A 6-45

(a) **Yes**

(b) **No** Calcium enters smooth muscle cells when they contract.

(c) **Yes**

(d) **No** It increases calcium absorption.

(e) **Yes** This is the value of ionized calcium concentration in plasma with which interstitial fluid is in equilibrium.

A 6-46

(a) **No** Ca^{2+} outside is many orders of magnitude greater than inside.

(b) **Yes**

(c) **Yes** This assists restoration of calcium level by aiding absorption.

(d) **Yes**

(e) **No** The requirement is twice as much—and the usual intake is nearer 1000 mg/day. 100 mg is the normal average value for net absorption.

A 6-47

(a) **Yes**

(b) **Yes**

(c) **No** -blasts are for 'building', -clasts are for breaking down.

(d) **Yes**

(e) **No** Calcitonin decreases the removal of calcium from bone.

A 6-48

(a) **No** All multiplication (mitosis) of ova is complete at birth: meiotic division occurs before ovulation.

(b) **Yes** Many Graafian follicles contain oocytes that develop during the first half of the cycle, reaching different stages; only one usually discharges at ovulation.

(c) **Yes** The developing follicles secrete oestradiol under the influence of FSH.

(d) **No** The progesterone which is secreted during the second half of the cycle has a negative feedback effect on LH (which maintains the corpus luteum and its secretions); it is oestrogens which have a negative feedback effect on FSH.

(e) **No** The Graafian follicle which has discharged its egg cell becomes a corpus luteum in the second half of the cycle; this is maintained into pregnancy if fertilization occurs.

A 6-49

(a) **Yes** The endometrium and its blood vessels proliferate under the influence of the increasing oestrogen secretion from the developing follicles in the ovary.

(b) **No** The second half of the cycle is the secretory phase, but it is mucus which is secreted; gonadotrophins start to be secreted by fetal tissue before the end of a cycle in which fertilization and implantation has occurred.

(c) **Yes** As the second half of the cycle progresses, the secretion of both the hormones by the corpus luteum rises, and then falls; the fall is followed by disintegration and shedding of the endometrium.

(d) **No** It becomes thinner at mid-cycle, facilitating the passage of sperms.

(e) **Yes** Body temperature rises at the time of ovulation, providing a clue for devotees of the rhythm method of contraception.

A 6-50

(a) **Yes** Secretion of gonadotrophins starts as soon as implantation has occurred and they take over from maternal hormones in maintaining the corpus luteum.

(b) **Yes** This causes enlargement and proliferation of the alveoli of mammary glands.

(c) **Yes**

(d) **No** Chorionic gonadotrophin, which is fetal, is the relevant hormone.

(e) **Yes** The cardiac output increases; the blood pressure does not increase; total peripheral resistance is therefore lower.

A 6-51

(a) **Yes** The development of a testis rather than an ovary is genetically determined, and the secretion of testosterone by that testis governs the further development of male characteristics.

(b) **Yes** Testosterone secretion influences both the fetal and pubertal development.

(c) **No** Androgen secretion is controlled by ACTH.

(d) **Yes** Luteinizing hormone controls testosterone secretion, otherwise known in the male as interstitial cell stimulating hormone.

(e) **No** Testosterone, as well as pituitary gonadotrophins, are necessary for spermatogenesis.

A 6-52

(a) **Yes**

(b) **No** It rises due to increased sympathetic activity.

(c) **Yes**

(d) **No** Centres are in the hypothalamus.

(e) **No** Shivering produces heat, which helps to raise temperature.

A 6-53

(a) **No** His metabolic rate is depressed.

(b) **No** The shivering reflex is paralysed.

(c) **Yes**

(d) **Yes**

(e) **No** Vasomotor tonus is absent.

A 6-54

(a) **Yes** Sweating is ineffective if you rub the sweat off with a towel.

(b) **No** Not in man, only in panting animals, which are furry and cannot use sweating.

(c) **No** Sweat contains NaCl; sodium loss can be a problem, and intake should be increased.

(d) **No** The minimum volume, of maximally concentrated urine is 400 to 500 ml/day.

(e) **Yes** Increased blood flow through the skin requires this.

A 6-55

(a) **No** It decreases; thyroid activity decreases.

(b) **No** The sweat glands are cholinergically innervated.

(c) **No** The central receptors are about 50 times more potent than the peripheral receptors.

(d) **No** A high relative humidity decreases evaporation from the body surface.

(e) **Yes** The set point of the hypothalamic integrating centre is set at a high level, so the subject feels cold until his core temperature is raised to match the new set point.

A 6-56

(a) **No** Insensible perspiration also provides fluid for evaporation.

(b) **Yes**

(c) **No** In a trained athlete the sodium concentration in sweat is less than half that in plasma.

(d) **No** They are in the hypothalamus.

(e) **Yes** At this temperature heat stroke develops and is commonly fatal.

Index

(*The numbers in this index are the question and answer numbers*)